Climate – our future?

Dedication

This book is dedicated to Professor Hans Oeschger, one of the pioneers in the science of global change.

Climate – our future?

Author and book concept

Ulrich Schotterer

Art director

Peter Andermatt

Translation

Kerry Kelts with Anne Arquit-Niederberger and Ulrich Schotterer

Technical contributions

Anne-Christine Clottu Vogel from the Swiss Academy of Sciences and Christof Blum of Kümmerly+Frey helped streamline all aspects of the production of the book

Color separations: Bufot GmbH, Basle, Kurt Bütschi

Typesetter: Elgra AG, Berne

Binding: Schlatter AG, Berne

Printed in Switzerland by Kümmerly+Frey, Geographical Publishers, Hallerstrasse 6–10, CH-3001 Berne

Science contributions

The chapter "Weather and climate" was designed and written with the help of Gian Gensler and Max Schüepp; Anne Arquit-Niederberger assisted in formulating the summaries at the end of each chapter; Thomas D. Potter provided a careful scientific review of the entire English manuscript; and the following scientists provided support through suggestions and discussions, or by providing data and photographs. The graphic illustrations are based on a number of scientific sources, primarily authors of the IPCC (Intergovernmental Panel on Climate Change) Report 1990.

Anne Arquit-Niederberger
Georg Budmiger
Fulvio Caccia
Hans-Ulrich Dütsch
Claus Fröhlich
Gerhard Furrer
Fritz Gassmann
Barbara Gerber
Walter Good
Georges Grosjean
Hanspeter Holzhauser
André Junod
Kerry Kelts
Bruno Messerli
Hans Oeschger
Christian Pfister
Thomas D. Potter
Fritz Schweingruber
Heinz Wanner
Urs Wiesmann
Matthias Winiger
Heinz Zumbühl

Artistic contributions

Graphics students from The School of Arts, Berne, transformed science into art and children from The International School of Berne drew their visions of the future. Their teachers, Peter Andermatt, Stefan Bundi, Bruno Cerf, Claude Kuhn, and Jean White, provided the necessary technical guidance, but allowed the students to create.

Eva Baumann
Verena Baumann
Elsi Brönnimann
Silvia Brühlhardt
Walter Buri
Catherine Eigenmann
Edith Helfer
Katja Leudolph
Lukas Machata
Sibylle von May
Roberto Renfer
Andreas Stettler
Agnes Weber
Karin Widmer

First English edition, 1990
First North American Edition, 1992
University of Minnesota Press,
2037 University Avenue Southeast,
Minneapolis, MN 55414

Originally published 1987 in Switzerland by Kümmerly+Frey under the auspices of the Swiss Academy of Sciences. The first English edition was sponsored by the Swiss government on the occasion of the Second World Climate Conference

Publication of the North American edition of this book was made possible by generous support from the Limnological Research Center and the Department of Geology and Geophysics, University of Minnesota, and by a bequest from Josiah H. Chase to honor his parents, Ellen Rankin Chase and Josiah Hook Chase, Minnesota territorial pioneers.

Library of Congress Cataloging-in-Publication Data

Klima – unsere Zukunft? English.
Climate – our future? / Ulrich Schotterer; art director, Peter Andermatt; translated by Kerry Kelts with Anne Arquit-Niederberger and Ulrich Schotterer.
p. cm.
Translation of: Klima – unsere Zukunft?
ISBN 0–8166–2130–6 (hc)
1. Climatic changes. 2. Man–Influence of climate. 3. Man–Influence on nature. I. Schotterer, Ulrich
QC981.8.C5K5313 1992
551.6–dc20 91–32584
CIP

Preface

It has been a long road from coauthoring the paper "A box-diffusion model to study the carbon cycle in nature" in 1974 to creating a picture book to stimulate thought about the Earth system among nonscientists. In 1986, the Swiss Academy of Sciences lent its support to the idea of making climate change understandable to a wider public by publishing a picture-based book.

The first version of the book was published in 1987 to inform a Swiss audience in three of their national languages: German, Italian and French. In November 1990, the Swiss government presented participants of the Second World Climate Conference with a gift of the completely revised English translation, *Climate – our future?*

Because of increasing pressure to transmit ideas by images rather than by words, we worked closely with students of graphic arts. Their illustrations reflect a number of scientific sources, mainly from authors who contributed to the 1990 Report of the IPCC, the Intergovernmental Panel on Climate Change.

Images tell the story, figure captions enhance the essentials. Text passages tie and highlight the story, rather than provide long explanations of complex scientific reasoning. Brief summaries at the end of each chapter review knowledge of which everybody should be aware. They are based on the scientific assessments of the 1990 IPCC Report.

The publication of the English edition would not have been possible without the help of Kerry Kelts, who provided a translation of what I intended to convey in my native German tongue, and the artistic ability of Peter Andermatt, teacher and artist, who designed the layout of the book with a feeling for how to convey scientific concepts through images.

Ulrich Schotterer
ProClim – National Institute for Climate and Global Change
Swiss Academy of Sciences

Foreword

Changing climate was once in the domain of fate. We now learn that the burning of fossil fuels and other human activities since the early nineteenth century – only a minute ago in geologic time – may override natural climate trends. This is a global problem. Increasingly, we perceive a mutual responsibility and are preparing our defense against this collective societal threat. This book arms the reader with art and state-of-the-art science to explore the front.

In November of 1990, the report of the Intergovernmental Panel on Climate Change was presented to the Second World Climate Conference. It detailed convincing evidence that a continuation of greenhouse gas emissions at present levels, notably of carbon dioxide, methane, nitrous oxide, tropospheric ozone and chlorofluorocarbons, can progressively increase global mean temperatures and likely affect global climates. Although severity, rate and regional aspects of climate change cannot yet be predicted with confidence, their potential impacts pose threats to both natural and socioeconomic systems of this planet.

Climate – our future? looks outward from the Alps, a cradle of meteorology, to the world at large. Can a North American audience be interested in a European perspective on the problems of climate change? We think so. The issues and processes are global. Change in Europe is relevant to North America. Informed citizens want multiple perspectives.

Swiss Alpine images illustrate general issues. Processes that drive the daily weather are valid elsewhere, although consequences differ. The Alpine Föhn has its relative in hot Santa Ana winds that descend from California mountains. Both North America and Europe experienced the impact of Ice Ages and the capricious switchover to modern climates. Each is affected by changes in sea surface temperature. Europeans deforested their lands in the Middle Ages and recorded catastrophes during a Little Ice Age when concurrent climate changes perhaps motivated Indian migrations in North America.

Although the book does not emphasize certain climate features of North America – East Coast hurricanes, midwestern tornadoes, cold fronts from Alaska, rising Great Salt Lake levels, Rocky Mountain rains or California droughts – basic issues remain the same. Whereas public debate focuses on whether the weather will get hotter or colder, a major concern is whether we will have more or less water. And here the snapshots from global climate models are fuzziest. Past climate trends teach us to expect threshold responses and surprises.

This is personal literature. You are not following a reporter's tale. You are asked to look at the climate story from six different angles: the physical system, past changes, how climate affects cultures, how people affect climate, the importance of research and how to view our future. Each page distills one theme using art to convey an essence of science. Images invite curiosity, humor and irony. The book is also packed with facts – in fact, from research through 1990. Yet, the author and artist paint from a mesmerizing palette that can be enjoyed at many levels – from curious children to climate experts; from coffee table to college textbook.

What does an image imply? How might it reflect your own local climate interactions? Climate change is not a simple concept reserved for scientists; and we should not expect absolute predictions. Issues are as pervasive, complex and individual as politics, economics and religion; good and bad are relative. Climate is a cloudy science for the future; change is certain. But how fast? How much? Where? Implicit with change is time; forward and backward. Danger to societies is measured by rate rather than amount. We learn that rapid change disrupts, destabilizes. Haves and have-nots both confront questions of suspicion, guilt and action. Perceptions become as powerful as facts.

Throughout history, fruitful progress occurs during times of profound change. Slow threats leave room for optimism. The challenge of climate change presents an opportunity: to discover Earth as a system, intensify international solidarity, and seek new levels of technical quality that match the fundamental equilibria of our planet. Our future.

Kerry Kelts
Limnological Research Center
University of Minnesota

Table of contents

1. Weather and climate

Page 8

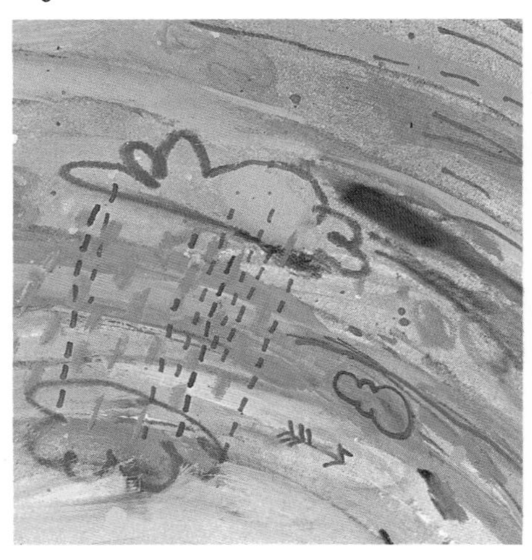

How to start; every day things to get your feet wet. What drives the weather in Switzerland and elsewhere? Extreme weather events. What can we expect from the future?

2. What we know about climate

Page 38

The thin, invisible veil of the atmosphere makes life possible and controls the temperature budget of the earth. Changes from ice age to warmer times are interrupted by sudden events. How stable is our climate system; could it switch over due to human interference?

3. Climate, humans and landscape

Page 74

Climate sculpts the landscape; climate and landscape imprint human society. Cultural heritage from the ice ages to the present industrial world. Are we the masters of our environment, or are we Nature's guests?

4. People – climate

Page 102

Population grows; the thirst for energy grows, greenhouse gases and pollution increase. The ozone hole. Deforestation, land use and the stability of natural ecosystems. Are we converting planet Earth into a better greenhouse?

5. Climate research

Page 128

Lessons from the past – anticipating the future. The global scientific effort. Is climate research providing a sound base for policy makers? International programs and conference activity. Are they steps from discussion to action? National programs for climate and global change, a Swiss example.

6. Climate – our future? A vision

Page 152

The world gets warmer – but will it be better? Sea level rises. Are traditional means fast enough? North-South inequalities? The farmer as landscape architect. Energy perspectives between dreams and reality. We reduce – and others?

1 Weather and climate

Climate – our future? That will mainly depend on how weather patterns change, because we are more perceptive to weather than climate. We use both terms matter-of-factly in our daily conversation and assume we clearly understand their meaning. Often, the opposite is the case. In the following pages, we aim to define these terms and add meaning to their usage with respect to our environment. Switzerland serves as our main model. The impact of man on climate has global character, but the consequences may be very different on a regional scale. We know today that the recipe for Swiss weather can be brewed as far away as the coast of Peru. We therefore open our view to the worldwide links between weather and climate.

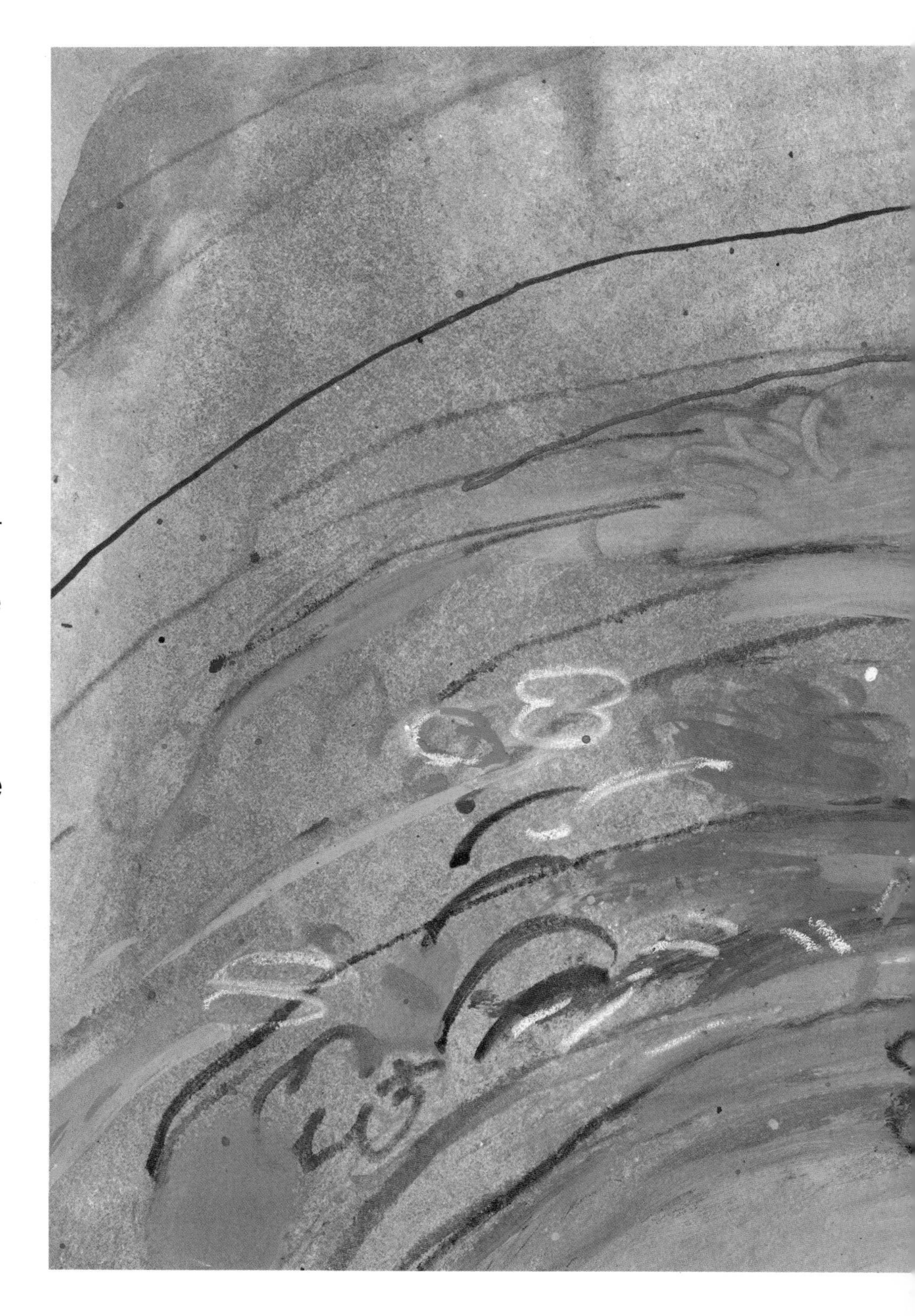

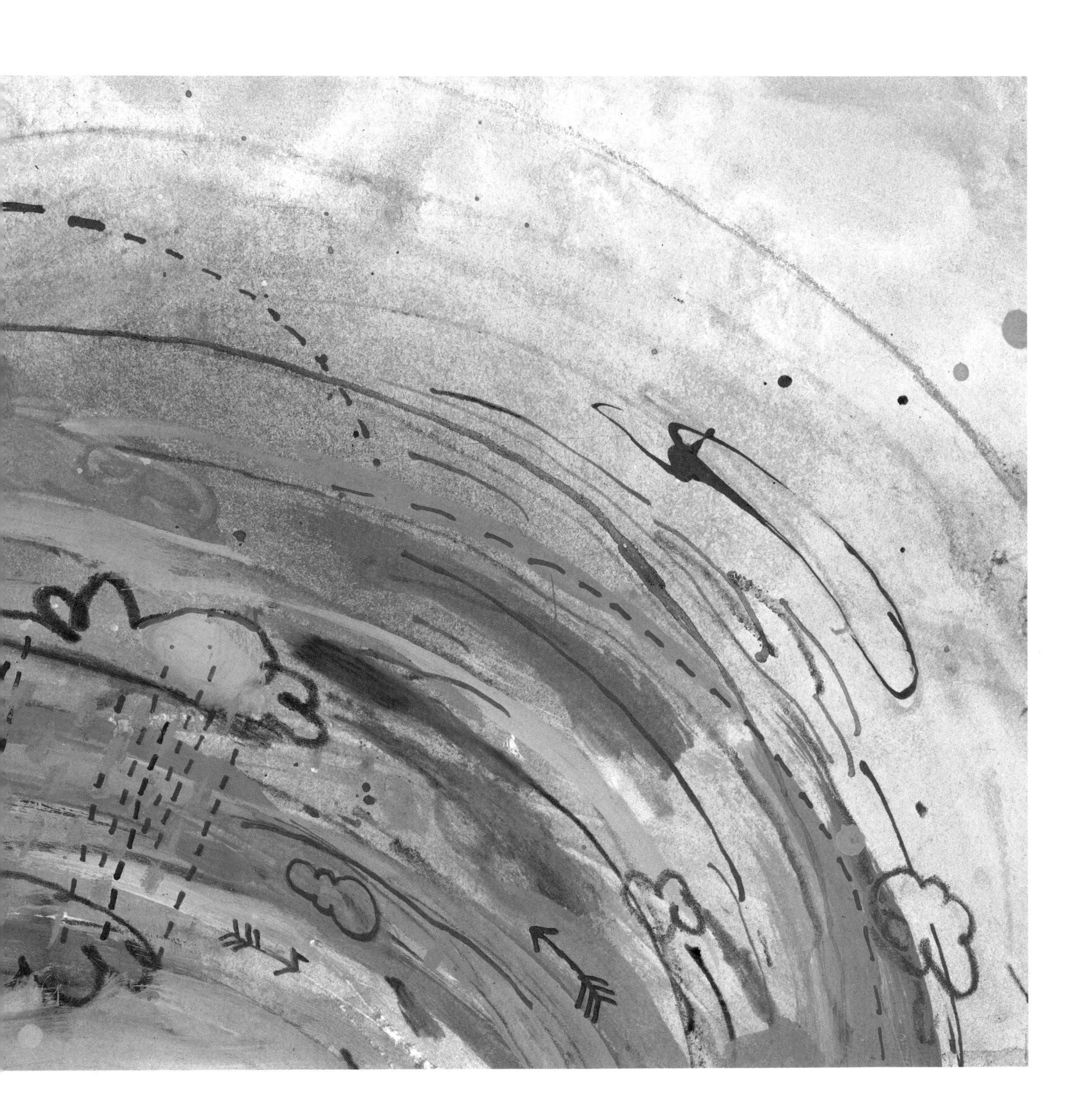

1 Introduction and definitions

Weather and *weather pattern* designate the momentary condition of our atmosphere. This snapshot is characterized by the values for various weather elements, such as temperature, air pressure, wind speed and direction, relative humidity, cloud cover, precipitation and visibility. The lifetime of a specific weather event is generally short. Often it lasts less than an hour, and rarely more than a few days. Humans react very sensitively to the precocious nature of weather and its extremes. *Climate* is the sum of all weather patterns in a certain region over spans of years to decades. The behavior of a single weather parameter, and its climatic implications, can be described statistically. Statistical analyses provide the basis for planning many activities that are weather dependent. A *weather pattern* provides a connection between weather and climate. It describes a dominant situation of several consecutive days, occasionally up to one or even two weeks. Every characteristic *pattern* for the alpine areas is directly related to the general weather patterns over Europe. Year after year, empirical observations tell us how the patterns are likely to produce local weather. The evolution and frequency of weather patterns remain more or less similar, and collectively over the years add up to our local climate. Near the Alps, regional-scale weather patterns can be strongly modified as air masses interact with topographic barriers.

1.1
A meteorologist's map for July 29, 1984, showing a summer high-pressure situation.

1.2
. . . and this is how the hiker might have experienced the weather on July 29, 1984, as he let his soul drift with the winds in the Swiss hills and dales.

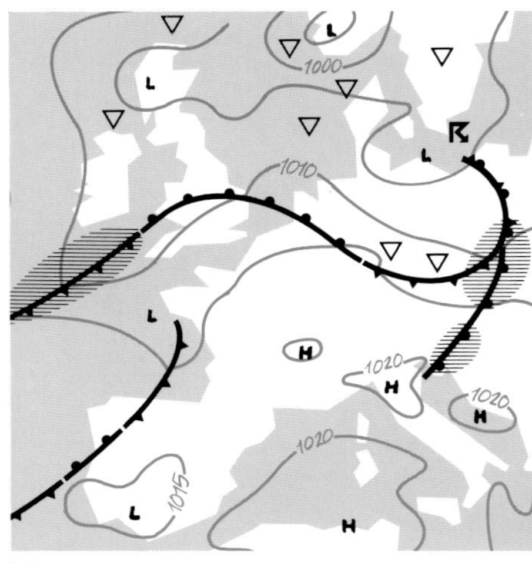

1.1

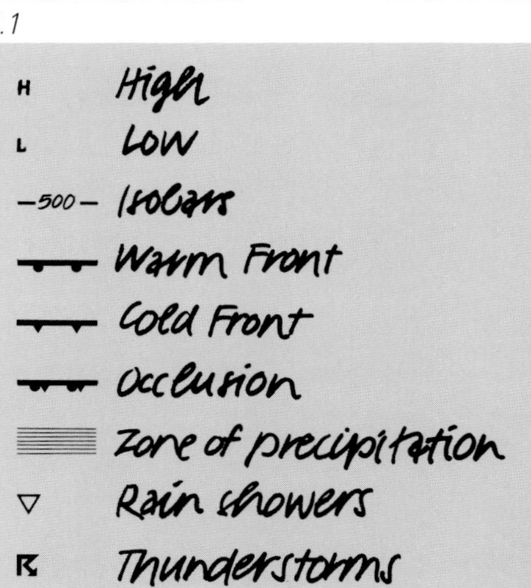

1.2

1

1.3
. . . and this is what the weather on midday July 29, 1984, looked to the European weather satellite METEOSAT from 36000 km altitude. This image is of the visible light spectrum and shows thick clouds in white; light-colored desert sands; medium shades for thin cloud cover or plant-covered land, and dark oceans.

Six typical weather patterns in Switzerland

We choose wind as the first criteria for classifying diversity. *Forecasts* of gentle breezes mean that the lines of equal air pressure – isobars – are widely spaced; whereas forecasts of strong, gusty winds mean steep pressure gradients. The first group – weak wind – thus encompasses high pressure values and low gradient patterns. The second group – strong winds –, includes Swiss weather conditions characterized by four main wind directions: Föhn, Westwind, North Pile-up, and Bise – cold dry winds from the northeast –. The first and second groups together form six typical weather patterns which are the ingredients for innumerable local weather situations.
A north- or southward shift in climate belts could alter the frequency of extreme weather events that create hardships in our environment.

1. High pressure

1.4
In the core of high-pressure zones air masses sink and clouds dissipate.
1.5
The white band of the snow-covered Alps as seen during a typically cloud-free, springtime high in central Europe.
1.6
Index map of names in Switzerland.
1.7
During winter highs, lowland valleys are under dense fog while snowy slopes of the Bernese Alps are bathed in cloudless Sunshine.

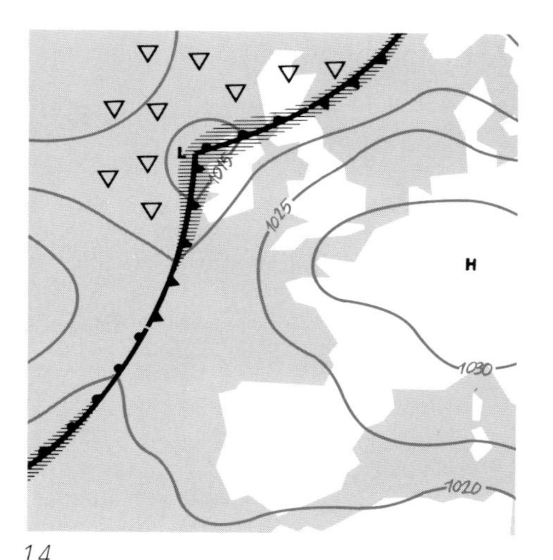

1.4

1.5

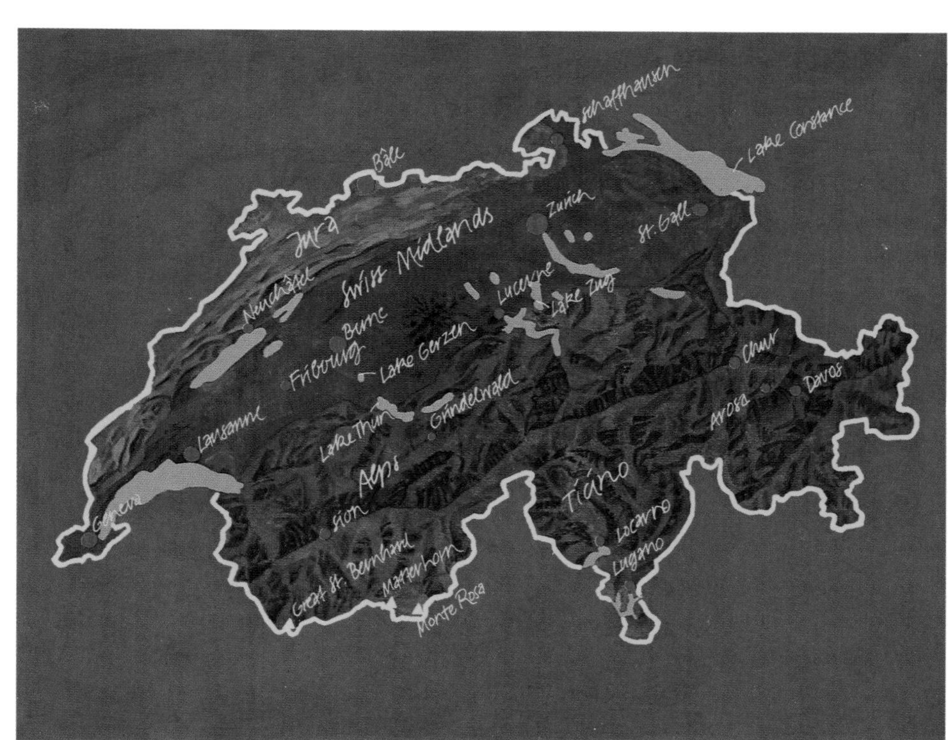

1.6

1.7

1

2. Low pressure

Temperature and pressure gradients decrease between polar and subtropical belts during the warm seasons when the poles receive continuous Sunshine. As a consequence, the pressure gradient over central Europe is often low during the summer. The warm continent heats the air masses which then rise, like soap bubbles, to higher altitudes. Huge thunderheads tower upwards as rising air masses cool, and water condenses. Temperatures to −40 °C produce ice crystals, and rain showers develop quickly. Strong thundershowers and gusts result, or perhaps hailstorms. The static electrical charge created in the atmosphere unloads with Nature's spectacle of lightning and thunder

1.8
Lines of equal air pressure – isobars – are widely spaced in Central Europe.

1.9
Summer thunderstorm over Lake Zug.

1.10
Thunderheads rise over the Monte Rosa Massif as seen from Colle Gnifetti plateau at 4500 m. A mobile laboratory in the foreground was used to collect automatic meteorological data during an ice-drilling campaign in 1982. Ice-cores contain meteorological information from the years before instrumental records.

3. Föhn

Föhn is German for a thermal wind in alpine valleys that is thought to cause headaches. During Föhn conditions, strong winds from the south or southwest cross the Alps. These carry moist air from the Gulf of Genoa to the southern foothills, where air rises and cools, producing rain. The cool air is then depleted of moisture. On the northern side of the Alps, these cool air masses sink rapidly down the valley slopes, accelerating and heating themselves by compression, arriving at the valley floor as a warm, dry airmass. Dry air provides the characteristic crystal clear visibility during a Föhn. In the lowlands the Föhn often collides with cold air masses occupying the valley floors. The interactions result in long, internal wavy patterns in the atmosphere with characteristically small, lens-shaped cloud trains called *lenticularis,* as shown in the drawing below.

1.11
Typical Föhn situation with a low in the west, high over Eastern Europe, and a nick in the front over the Po Plain, Italy.

1.12
A cross section of air circulation during Föhn conditions along a trace, Ticino–Alps–Swiss Midlands.

1.9

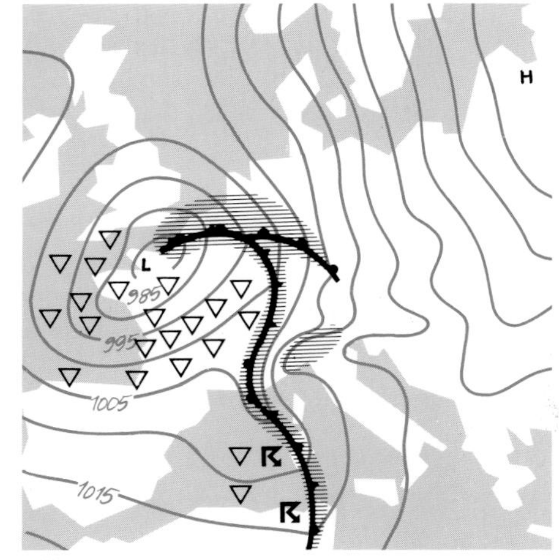

1.11

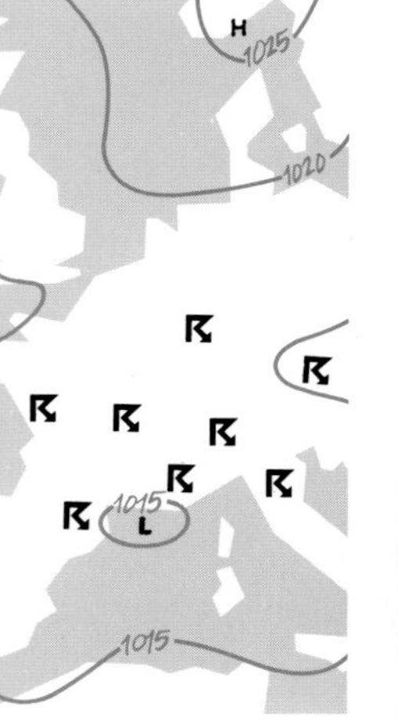

1.8

1.10

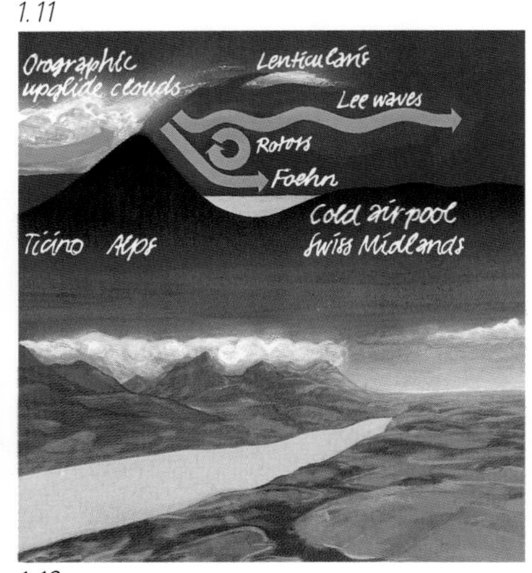

1.12

4. West wind weather

A well-developed low moves quickly from the Atlantic over the British Isles toward the North Sea. The elongated zones of air convergence – *fronts* – pass over Europe with thick cloud bands.

1.13
On November 8, 1982, a powerful Föhn storm moved down this forest near Lake Zug. Alpine valleys channel the Föhn, increasing its destructive impact.
1.14
Meteorological situation during west-wind weather. Compression of the isobars – stronger gradients – in the warm sector south of a low pressure center. The forecast includes cooling with wind directions changing from west to northwest on the trailing side, west of the low.
1.15
Strong November westerlies whip up foam across Lake Zug.
1.16
As the polar front moves towards the East, a warm air mass glides up over sluggish, cold air masses, cools, and causes thick cloud cover and precipitation over western and central Europe.

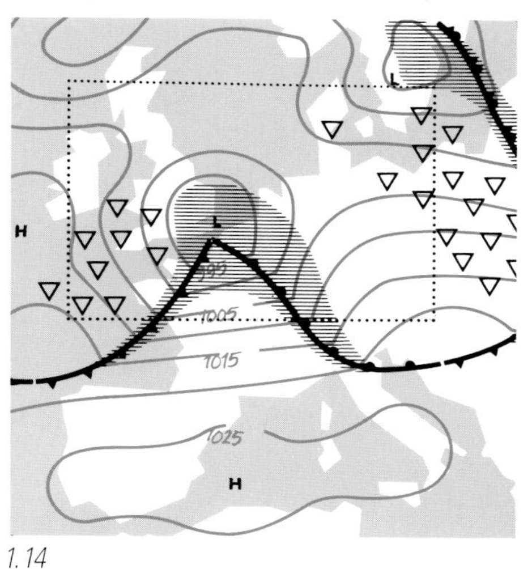

1.14

1.15

1.13

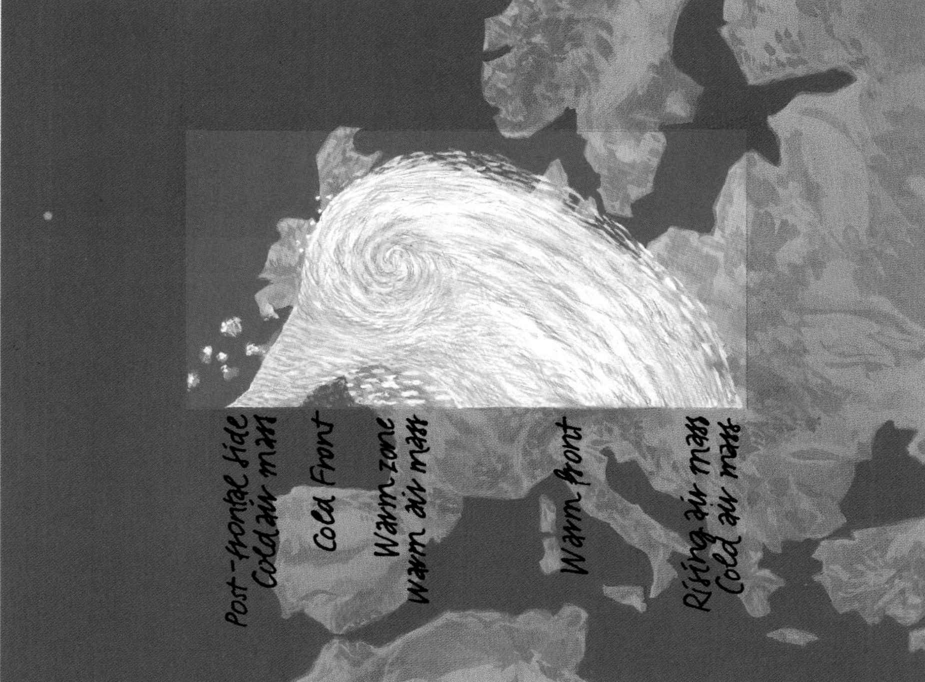

1.16

1

5. Pile-ups

Cold, dense air masses from the North Sea pile up against the alpine slopes as they move south, causing a thick cloud cover and heavy precipitation. Cold air flowing into the southern alpine valleys is warmed, resulting in North Föhn.

1.17
Air pressure conditions for Pile-ups contrast with those for a Föhn pattern.

1.18
Climbing Mt. Gridone, west of Locarno in the southern Alps during a North Föhn situation. Clouds visible to the north, flowing over the alpine crest, dissipate quickly.

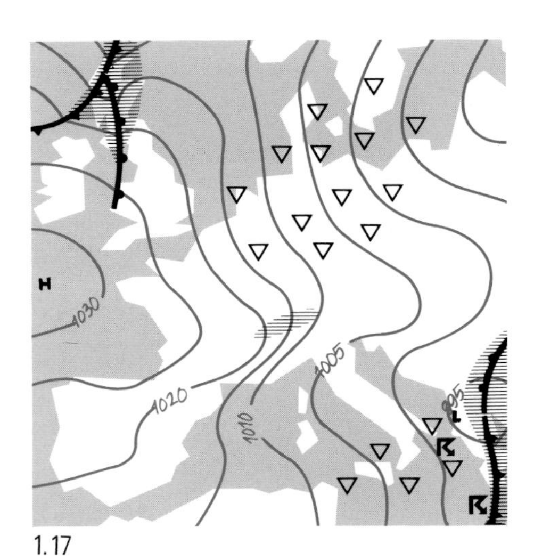

1.17

1.18

6. Bise conditions

Cold, northeasterly winds in Switzerland are called Bise winds. They flow along the midland corridor between the Jura Mountains and the Alps, becoming stronger as they are funneled towards Geneva. There, the Bise brings bitter cold, precipitation, and stormy, high seas on Lake Geneva.

1.19
Palm tree near Locarno, and a sky of fish-shaped clouds 'makerel sky' caused by a Föhn from the North. The alpine arc protects the southern valleys from the brunt of northern cold fronts and allows subtropical vegetation to flourish.

1.20
High pressure zone over the British Isles and northern Germany, and a low pressure center over the central Mediterranean Sea. The forecast: cold Siberian winter winds from eastern Europe!

1.21
A dock railing in Geneva during a winter Bise.

1.22
The air flow pathway of the northeasterly Bise across the Swiss Midlands.

1.19

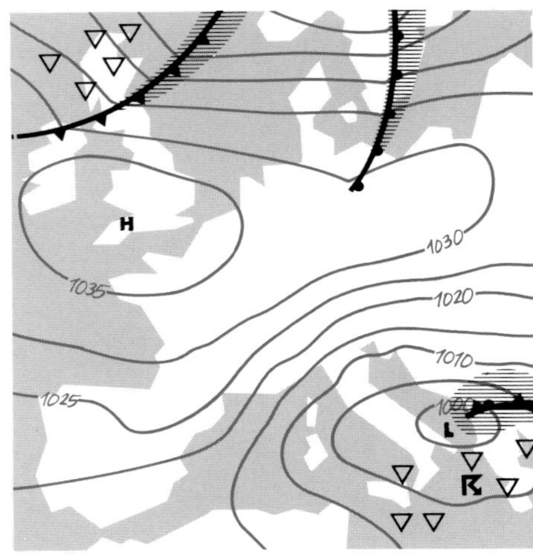

1.20

1.21

1.22

Flow pathways for the weather patterns 3–6

Periods of strong winds in the Alps plot closely spaced contours of equal air pressure. The relative position of high and low pressure centers determines the direction of wind. We need to remember only one rule: *air flows clockwise* out of a *high-pressure center and spirals anticlockwise towards a low-pressure center, where it then rises.* Pile-ups and clouds form along the windward side, the *luv,* of a mountain range; whereas Föhn-related clearing occurs on the side away from the wind, the *lee.* Weather watching is everyone's hobby!

1.23
Air circulation patterns between a high- and low-pressure region.
1.24
Map of Switzerland with major wind pathways. The numbers refer to :

3. Föhn Forecast. Warm Mediterranean air crosses the Alps.
4. West Wind Weather. Alternating warm and cold oceanic air masses.
5. Pile-Up Forecast. Cold air masses from the north collide with the Alps. Warm air blows as a North Föhn in the southern valleys.
6. Bise Forecast. Cold continental air masses flow from Eastern Europe, particularly in winter.

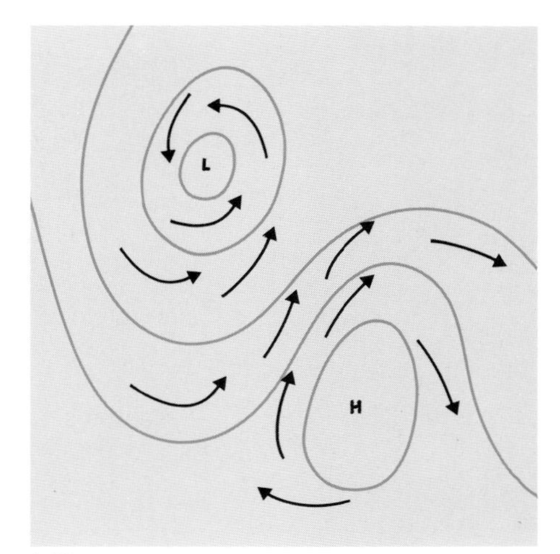

1.23

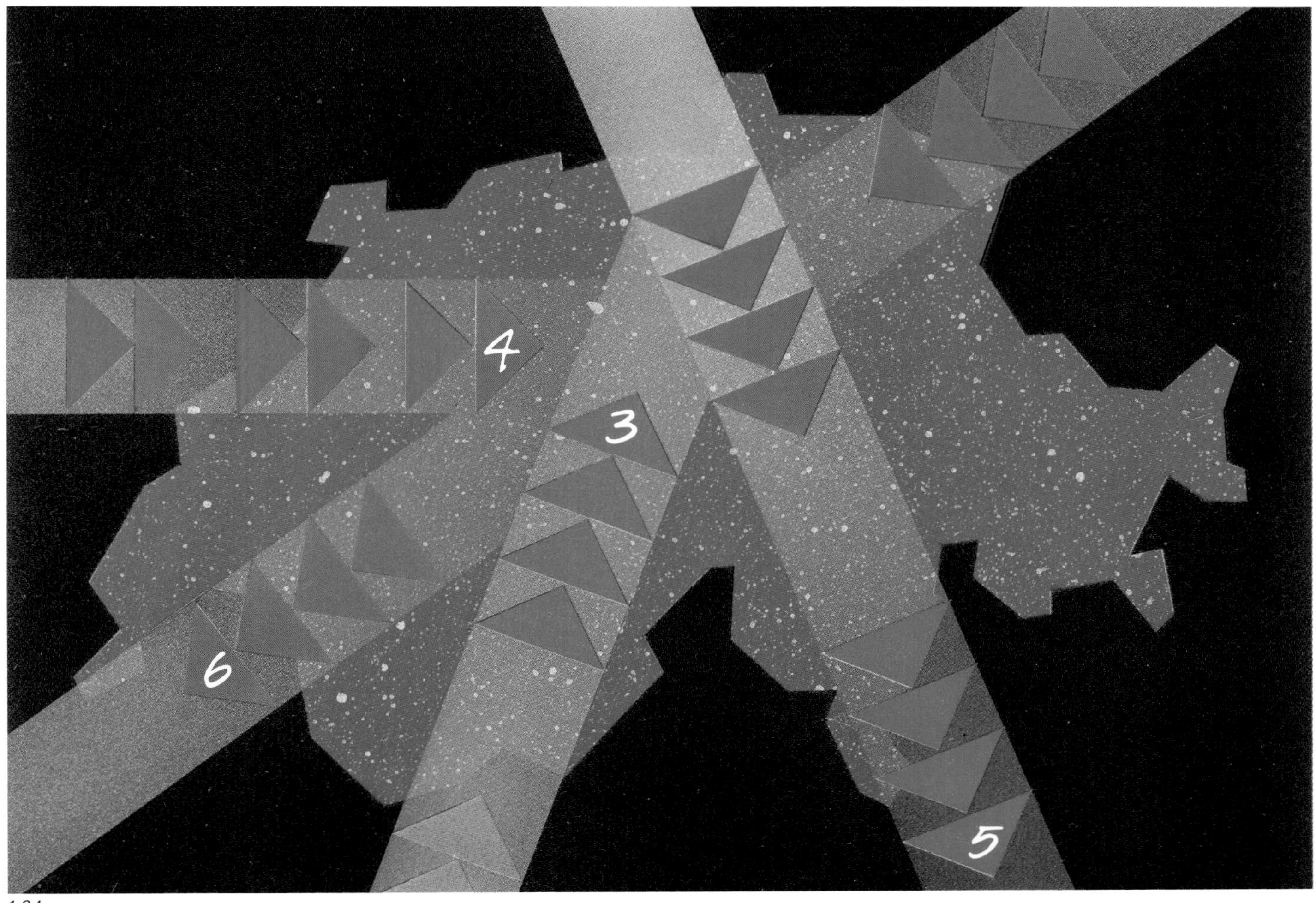

1.24

1 Weather extremes in the Alps

Climate variations are best felt during weather extremes, especially temperature and precipitation. Bâle set the record for the highest Swiss temperature with a value of 39°C in July, 1952. Centovalli, Ticino does the honors for highest rainfall with 400 mm on. September 10, 1983. The sum of whole periods can signal danger, such as in April 1986 when the Canton of Ticino was blessed with seven times the normal precipitation. A series of disastrous avalanches followed.
Long, dry intervals, up to 3 months, are common in Ticino in winter, and throughout the year

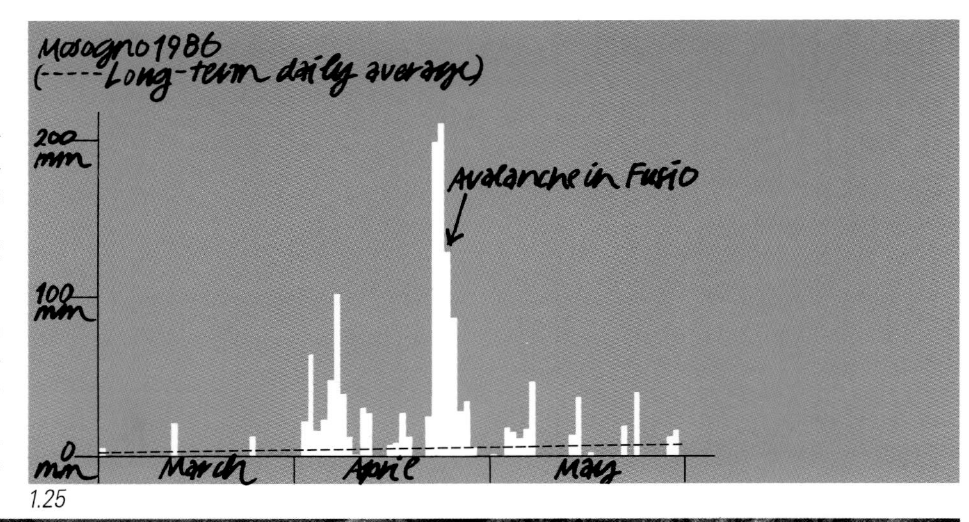

1.25

1.26

around the Mediterranean. These are responsible for dangerous forest fire conditions. If the extremes increase due to climate change, these sensitive regions will become further endangered.

1.25
Precipitation values for Mosogno, Ticino in the spring of 1986. Avalanches in April.

1.26
Air photo of an avalanche in Fusio, Ticino.

1.27
Precipitation pattern for Locarno in autumn 1983. Forest fires in December.

1.28
Forest fires near Locarno caused by extended dry periods.

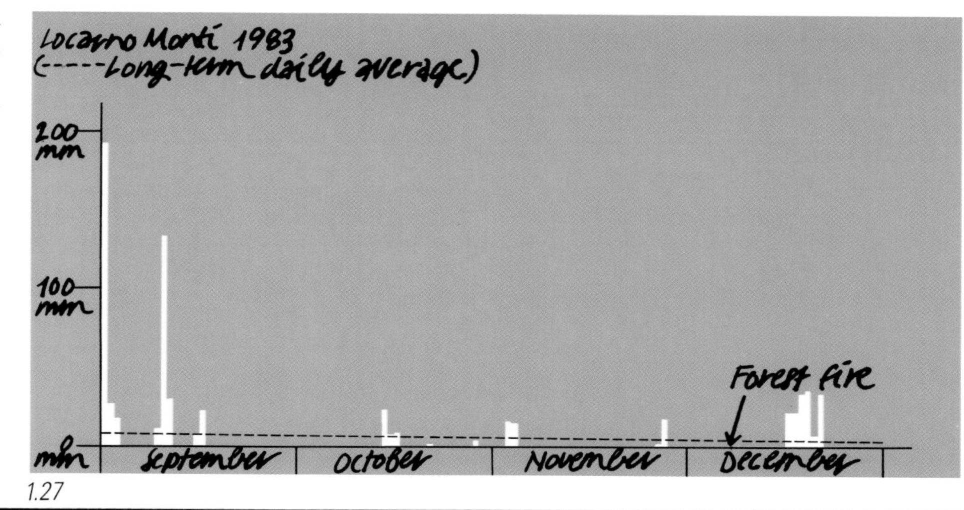

1.27

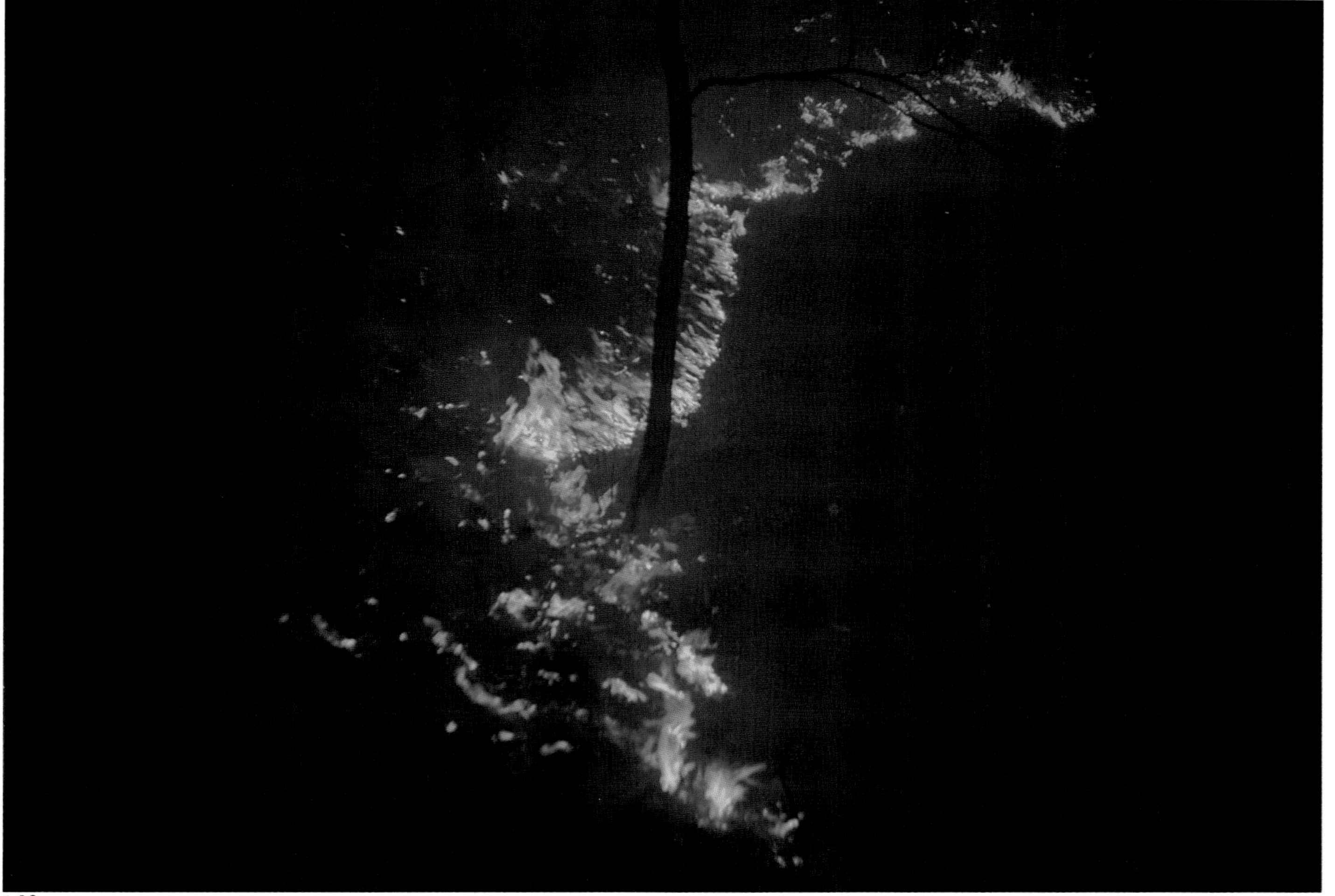

1.28

1 Weather extremes beyond the alpine realm

Unusual storms occur most commonly in the temperate zones during the winter. The western coastlines of Europe are hardest hit because these storms originate to the west or northwest. The historical data set suggests that major storm floods along the North Sea coastline have become more abundant in the 20th century. Although measuring criteria have changed, the years of extremes speak for themselves: major floods occurred in 1164, 1219, 1287, 1362, 1421, 1634, 1717, 1825, 1953, 1967, 1976, 1981, and in 1990.
Today, low-lying coastal belts can only be protected with enormous investments in technical barriers. Before, populations had to accept the loss of great strips of land.

1.30
A dyke is overflowed by northwest storms of long duration that pile water into the northern German coastal bays. Inhabitants behind the dykes live in constant fear of possible dam failures.

1.30

1.29
The sea reclaims its land.
Maps for the years 1240, 1634, and 1832 showing the invasion of the sea into North Friesen along the German North Sea coastline.

1 Tropical cyclones

Tropical cyclones can only form above ocean waters with sea surface temperatures – SST – in excess of 25 °C and in air masses with sufficient humidity. High water temperatures drive strong thermics and increased evaporation which create high-altitude, dense clouds and hurricane-strength winds. These storms may last up to several weeks, as long as the extreme low-pressure eye remains over the open ocean. They thus survive long enough to be baptized with names, once female, now mixed. As soon as tropical cyclones move onshore, they attain infamous stature due to their destructive power.

1.31
A very unusual cyclone formed in the Mediterranean Sea north of the Syrte Bay on January 25, 1982. In spite of a warm SST, this region usually lacks adequate air humidity due to the dry northeastern Trade Winds. Both are necessary conditions for cyclone formation.

1.32
Impression of a tropical cyclone. Most of these storms weaken or dissipate before causing major damage.

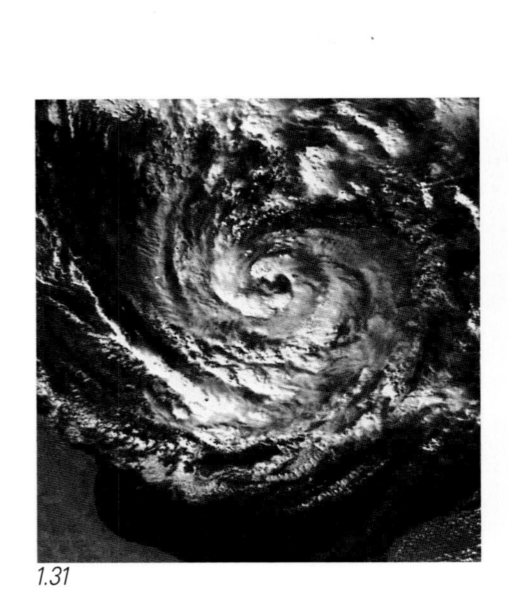

1.31

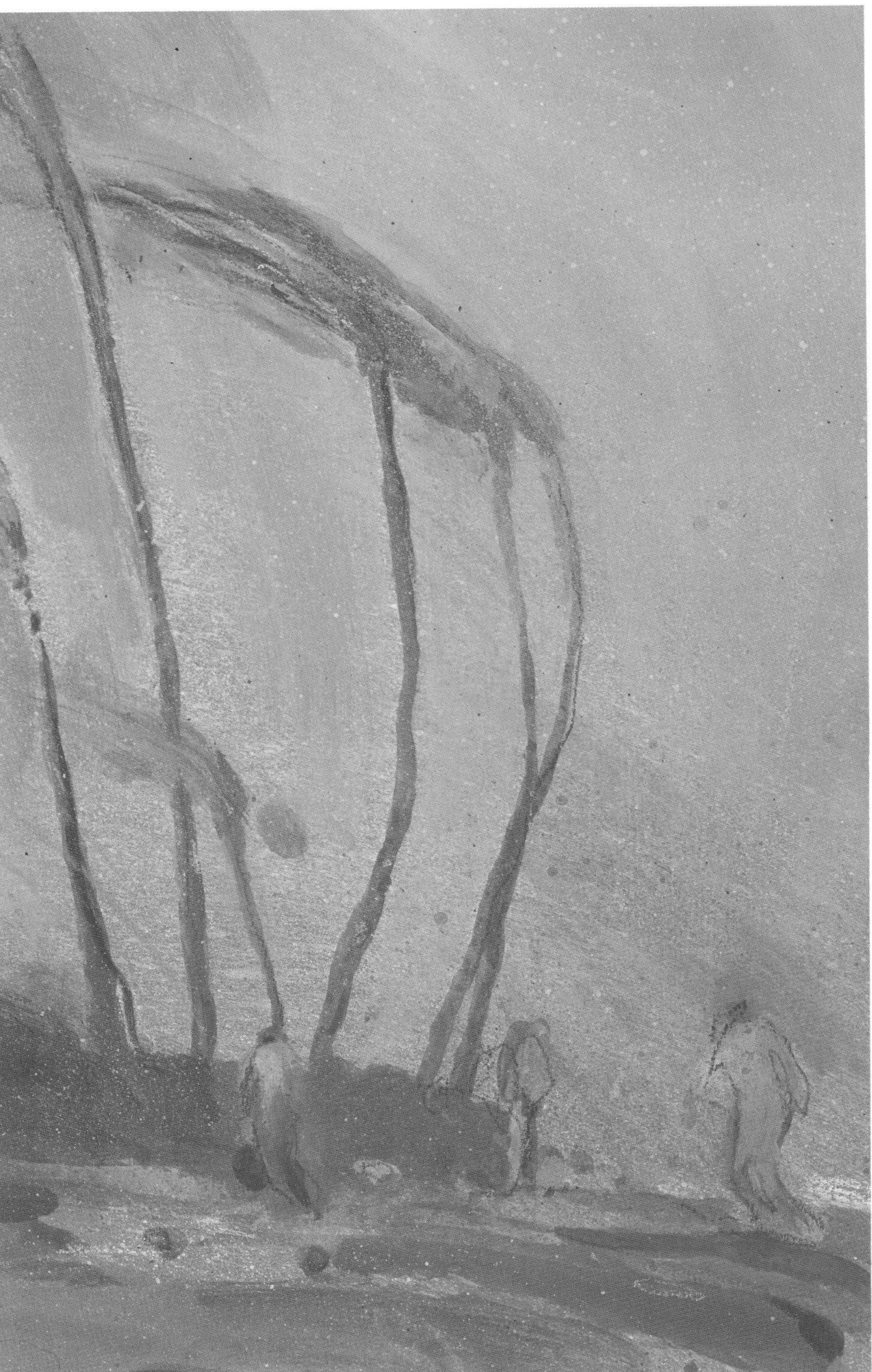

1.32

Sandstorms

Certain air pressure patterns, such as a low-pressure center between Spain and Morocco, may cause huge warm air masses to explode out of the Sahara toward the north. If rising thermics are strong enough to lift the massive dust at least 5000 m high, the dust takes a direct path to the Alps. Twenty-four hours later, the winter snow is dusted reddish to yellow. Most of the Saharan dust, however, is spread over the Atlantic by the Trade Winds and some may reach the Antilles or Florida after a cruise of 5000 km. Estimates from satellite images indicate that the Sahara delivers over 200 million tons of dust to the Atlantic in a single summer. This is more than the dust output of 10 major volcanic explosions.

1.33
Satellite image from July 28, 1983, showing the pathway – accentuated in yellow – of a dust cloud moving from the northern Sahara, across the Mediterranean and toward the Alps.

1.34
Impressions of an ominous sandstorm engulfing the sky.

1.33

1.34

1

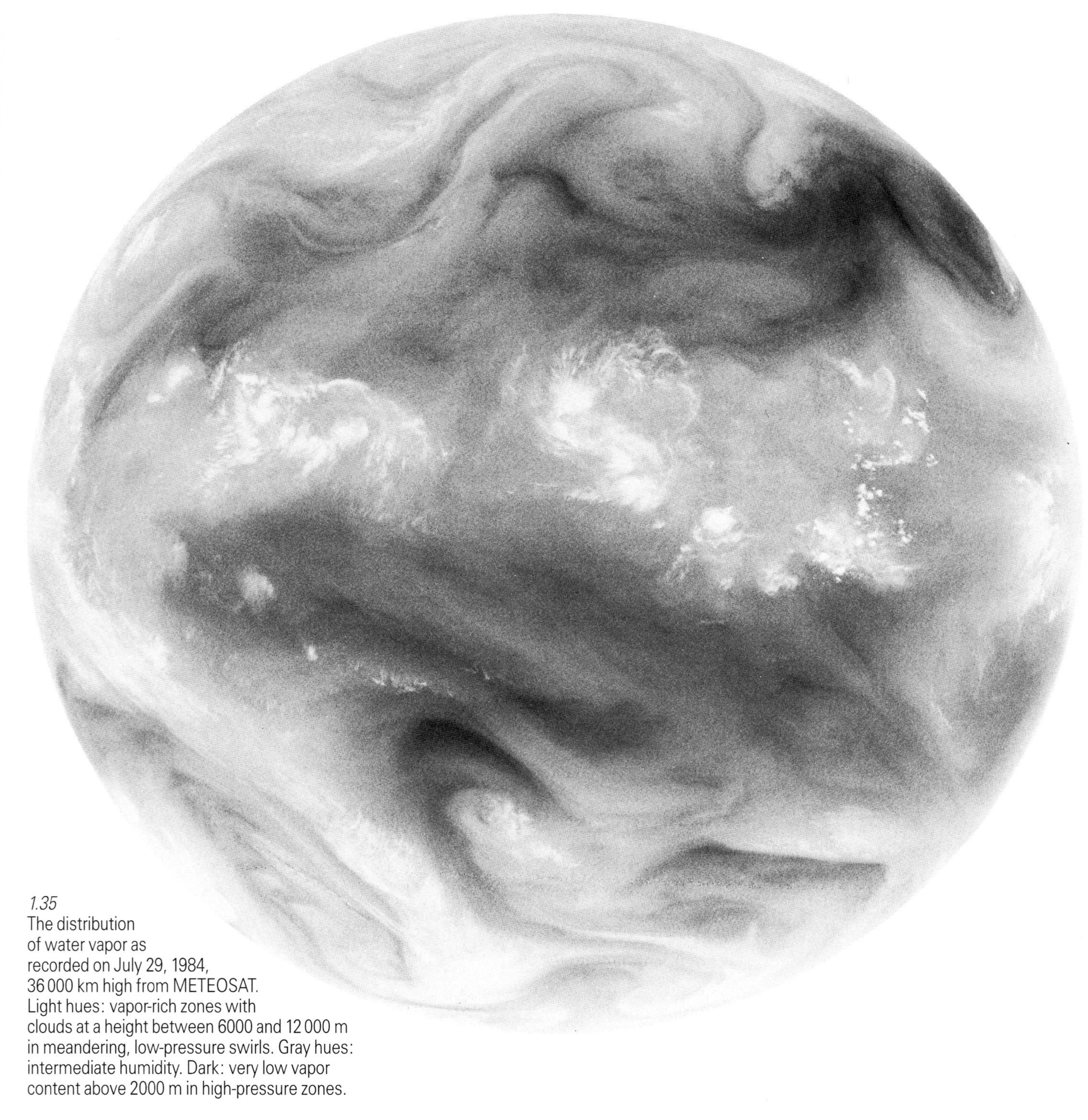

1.35
The distribution
of water vapor as
recorded on July 29, 1984,
36 000 km high from METEOSAT.
Light hues: vapor-rich zones with
clouds at a height between 6000 and 12 000 m
in meandering, low-pressure swirls. Gray hues:
intermediate humidity. Dark: very low vapor
content above 2000 m in high-pressure zones.

Clouds, vapor and wind

These three climate parameters are closely related. Clouds only form under certain weather and climate conditions, which depend on local air pressures, thus winds. The average cloud cover follows characteristic patterns in many regions of the Earth. One typical example is the rhythmic north and southward shifts of the Intertropical Convergence Zone – ITCZ. This marks the line of interference between humid equatorial monsoons and dry trade winds. Along with solar insolation, cloud cover and atmospheric water vapor determine climatic zones. Water vapor is invisible to our human eye, but not to satellite sensors. In swirls and eddys resembling marble fudge ice cream, vapor masses are exchanged between the hemispheres as a sign of energy transfer. In this manner, events in the southern Pacific can have an impact on North America, or even Europe.

1.36
The band of cloud cover across Africa shifts with the equinoxes. Clouds are linked with the ITCZ and in summer months, barely extend north to the thirsty Sahel Zone at 10°–15° N to drop a load of very unpredictable rainfall.

1.37
During the southern hemispheric summer, the ITCZ moves south toward the equator and leaves the tropical forests free of cloud cover.

1.38
Bad weather signs over Greenland! Cloud cover is streaked out as a result of warm air masses that glide over colder air.

1.39
In the tropics, as here along the Rift Valley scarp of Kenya, thunderheads form as plumes of heated air masses rapidly rise.

1.36

1.38

1.37

1.39

1

1.40
Weather vane for central Europe. The length of bars in this illustration show the probable speed and direction of surface winds. The numbers indicate the relative frequency of times with very gentle breezes, given in percent. The Feldberg region "10", in the southern Black Forest is a typical case that mirrors the mid-European norm of dominant SW or W winds. Alpine valleys, 52 and 6, channel winds into two modes.
1.41
The normal pattern of shifts in the ITCZ and barometric centers over India; for January on the left, and July on the right. The migration of the ITCZ northwards in summer spearheads the arrival of monsoon rains. The switchover from dominant low to high-pressure conditions over Asia results in the reversal from dry, northeasterly Trade Winds in winter to moisture-laden SW winds in summer.
1.42
A rosette showing the average yearly wind directions for dry and rainy seasons in the Sahel Zone. Note the symmetrical pattern which reverses dominant modes between northeast and southwest.
1.43
The global average barometric situation is given in light blue contours, in *hecto-pascal units,* formerly millibars, with respect to sea level. These determine dominant Trade Winds – in green arrows –, and the climate belts. Subtropical high-pressure cells extend along both sides of the equator. Extended lows hover further polewards along the subpolar zone. European weather is mostly controlled by the Azoran High and the Icelandic Low, known to every European school kid. Both hemispheres experience a broad west wind belt along mid-latitudes due to the general distribution of high and low pressure cells. A weak low-pressure zone extends along the equator and causes trade winds from both hemispheres to flow from eastern sectors towards the Equatorial Low. Monsoon winds are a special condition with abrupt winter – blue arrows – and summer – red arrows – switchovers, which are related to continental effects.
1.44
Peasant agriculture in semi-arid areas north-west of Mt. Kenya during the rainy season . . .
1.45
. . . and during the dry season.

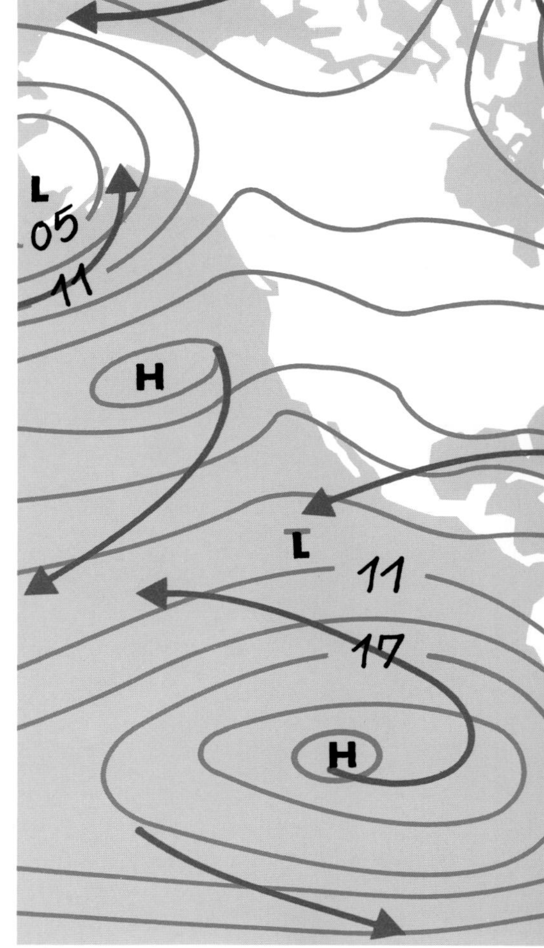

1.43

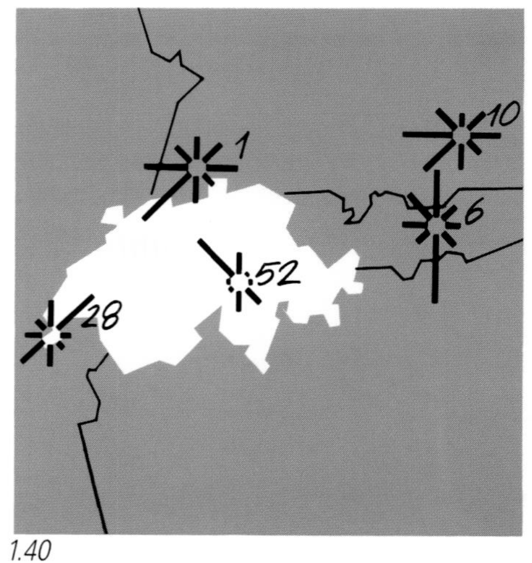

1.40

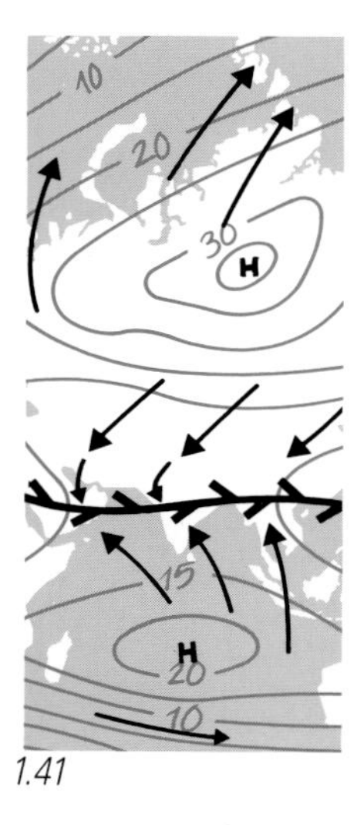

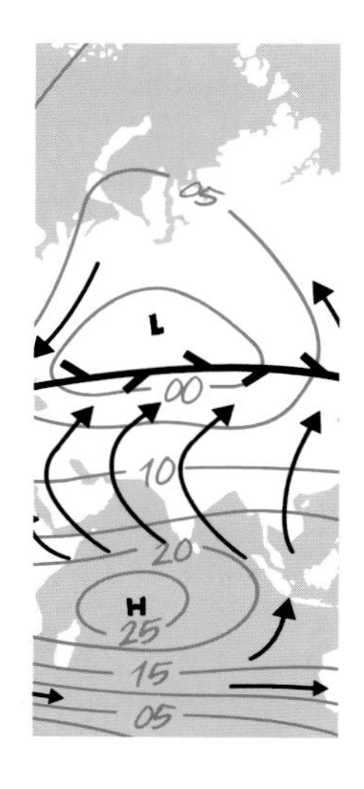

1.41

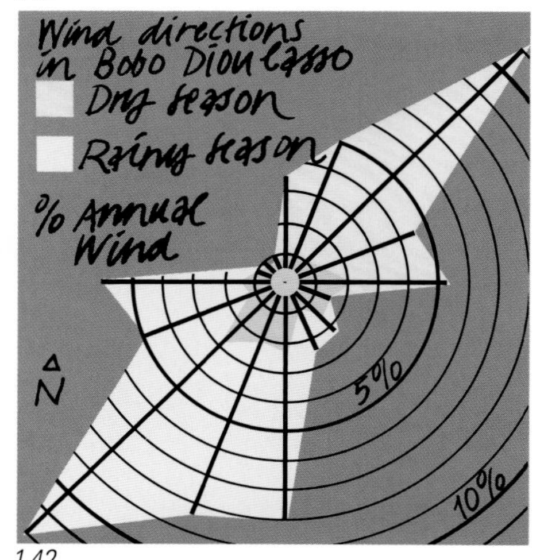

1.42

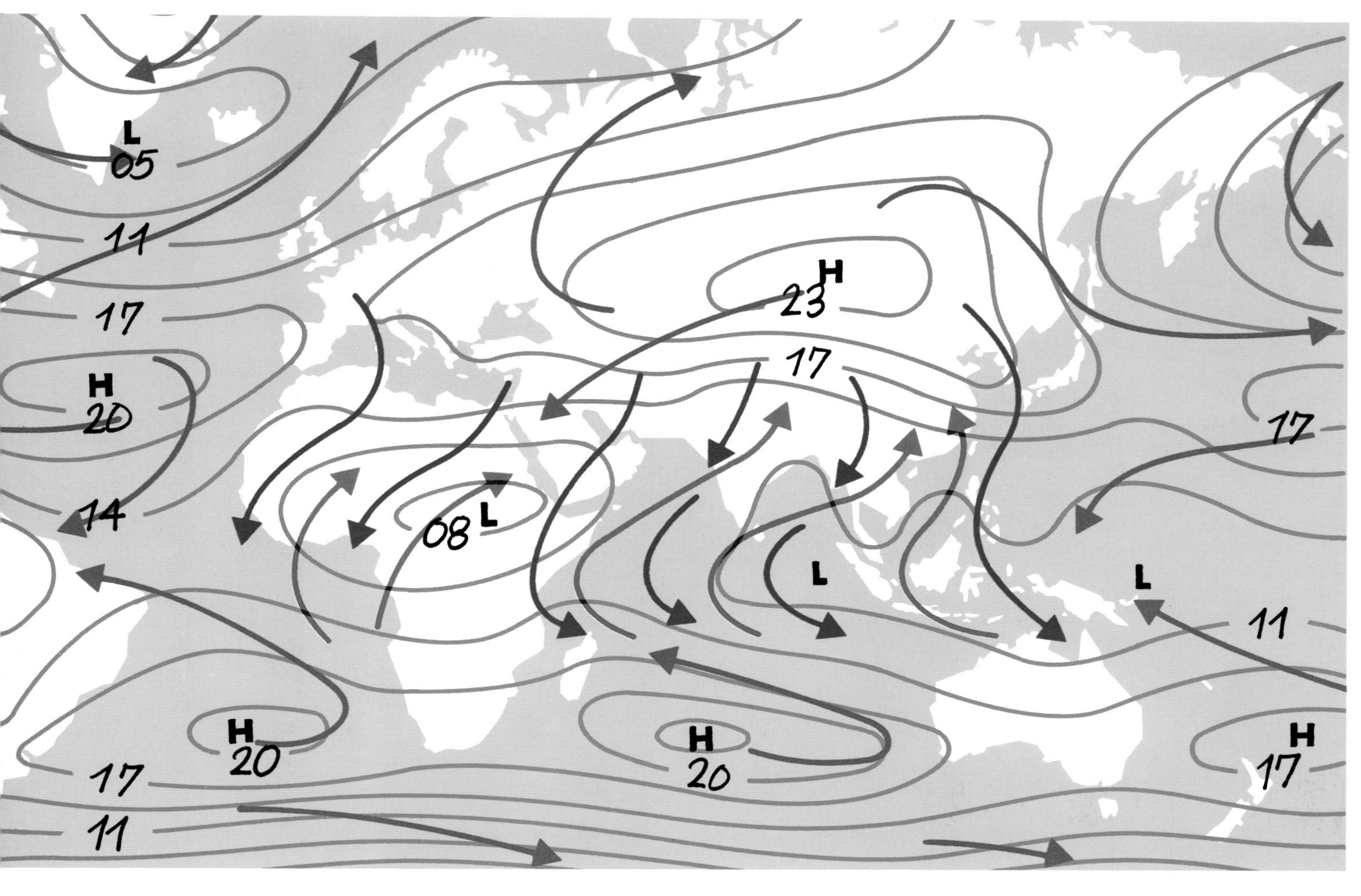

1.44

1.45

1

Sahel…

Two special anomalies of the climate system have been making frequent headlines in recent years. The *Sahel* attained sad notoriety due to frequent drought; the *El Niño* phenomena gained a mystical, distant aura. Both of these occupy us throughout this book because they are key mysteries illustrating our vulnerability to climate change. The Sahel Zone is a strip of land along the southern margin of the Sahara desert. Rains can only reach this outpost when winds have changed from northern Trade Winds to moist southerlies. The amount of rain decreases by ⅓ along a south to north transect of only 500 km – the distance from Paris to London. Combined with a 29 °C average annual temperature, this gradient defines a narrow, delicate transition zone from lush green savanna to arid desert. Highly variable summer rainfall leads to overgrazing during drier years. This promotes the encroachment of desert soil conditions towards the equator.

1.46
Distribution of average yearly rainfall – given in millimeter values – across west central African vegetation belts.

1.47
Life – as here in Niger – continues at the borderline of existence.

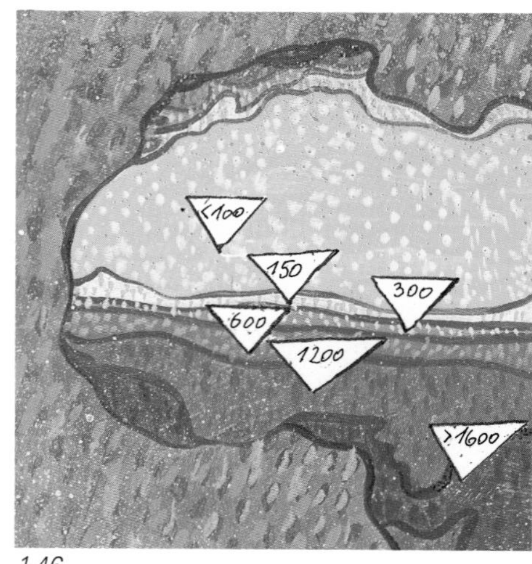

1.46

1.47

...and el Niño

In the equatorial region along the west coast of South America, large scale pressure oscillations influence winds. These drive the upwelling of cool, nutrient-rich waters from deeper Pacific Ocean levels. Upwelling is essential for a long, critical food-chain link from plankton to fish and birds. One end product is massive phosphate deposits from bird guano, important for the world fertilizer economy. During irregular intervals that recur between three and seven years, the dominant winds change to westerlies, which replace upwelling by nutrient-poor surface water masses. The food chain is clipped. Fish supply diminishes, fishermen seek new fishing grounds, and have thus caused international conflicts over fishing in the past. As an uncontrollable consequence, warmer surface water masses lead to increased evaporation and transfer of moisture that later may return as biblical flooding.

1.48
Graphical representation of the El Niño phenomena. The upper diagram shows normal conditions; cool, nutrient-rich, deep-water, in dark green, upwells due to winds. Below, the situation in an El Niño year when onshore winds keep a warm, equatorial lid on upwelling.

1.49
Life on the Galapagos Islands is adapted optimally to climate conditions

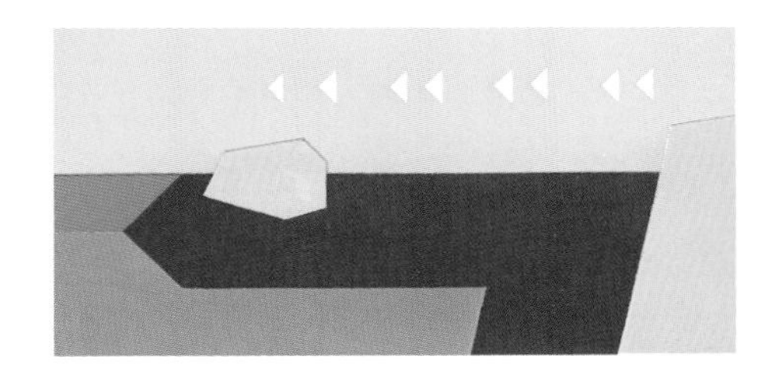

1.48

1.49

1 The snowline

The *snowline* refers to an altitude girdle above which the winter snow cover does not melt away during a subsequent summer. The snowline is related to latitude and is controlled by temperature and amounts of precipitation. The great polar ice caps persist because they receive limited solar radiation, which is largely reflected by their high *albedo*. These cold masses activate extended, subpolar low-pressure cells. Growth of ice, therefore, shifts the position of low-pressure belts, and thus affects regional climate. Global warming will reduce the extent of ice cover. This mechanism may contain a dangerous lever for pushing our usual weather patterns over a threshold. The extent of glacier ice and firn in the mountains of middle latitudes is also important. Snowy peaks along the equator determine local circulation and water budgets, and provide breathtaking tourist attractions.

1.50
Approximate level of the snowline along a pole-to-equator-transect. The snowline does not rise to its highest at the equator because of the zonally differentiated precipitation patterns.

1.51
Summit, the highest point in Greenland. Heaven and ice meet without contour over this vast emptiness. This location is now a crossroads for the international scientific community attempting to drill 3 km through the ice sheet.

1.52
The Findelen Glacier in the Swiss Alps in autumn 1983. Winter snow cover has melted back to about 3000 m elevation.

1.53
Mt. Kenya sits exactly on the equator. Numerous small glaciers persist among its craggy peaks at altitudes between 4800 and 5200 m. These temperate glaciers have retreated considerably during the 20th century.

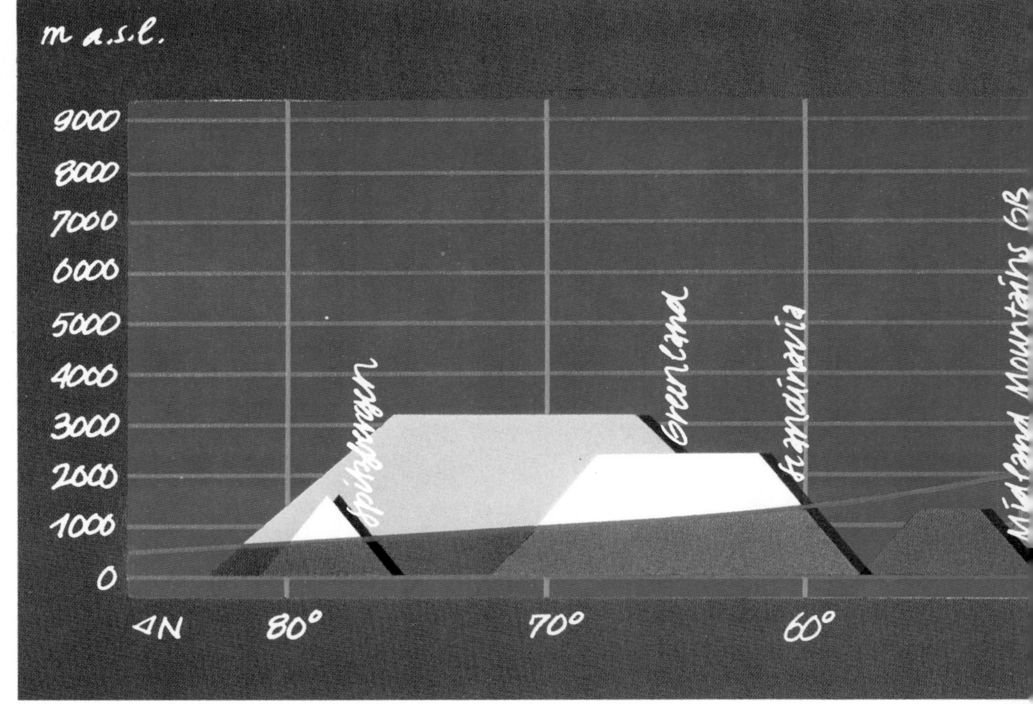

1.50

1.51

1.52

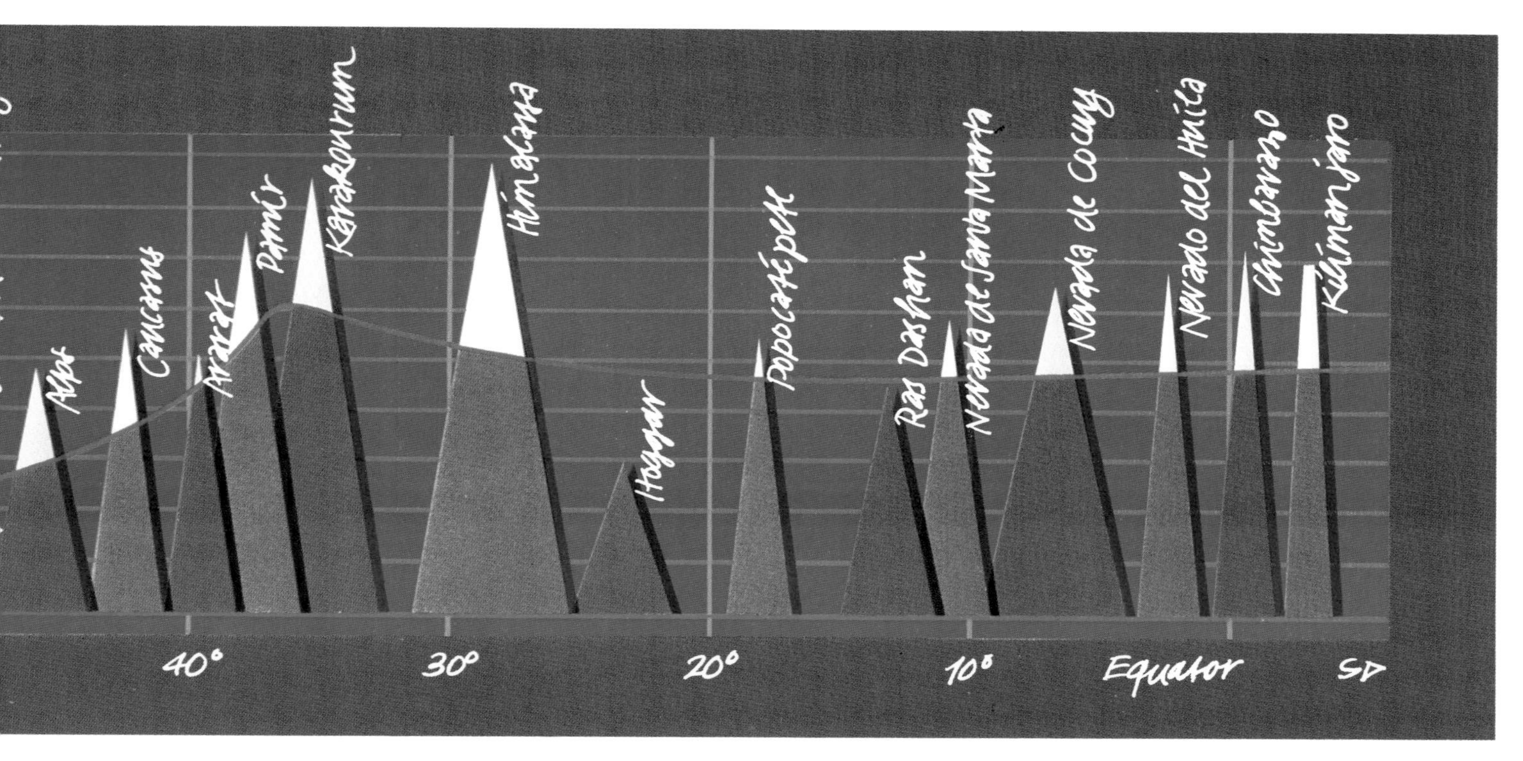
Alps
Caucasus
Ararat
Pamir
Karakorum
Himalaya
Hoggar
Popocatépetl
Ras Dashan
Nevada de Santa Marta
Nevada de Cocuy
Nevado del Huila
Chimborazo
Kilimanjaro
40°
30°
20°
10°
Equator

1.53

1 Climate pendulum The Swiss case study: Temperature

In the temperate zones, seasonal weather pattern changes are less dramatic than along the boundary zones between subtropics and tropics. The annual Swiss temperature curve displays a smooth bell. Characteristic anomalies, called *singularities,* occur at times. These accompany seasonal switchovers of the dominant supra-regional weather patterns. Typical examples include a Christmas Thaw, the Ice Saint in early May, a Sheep's Wool Cold Snap in mid June, and the Indian Summer in October. If the average temperatures in Zurich are plotted and we compare the differences for the period 1946–1985 with respect to the period for 1901–1945, certain patterns emerge. The Christmas Thaw has been less intensive during the last 40 years than before. February, however, has begun much warmer, with a colder end by 2½ °C. The curve doesn't show a Sheep's Wool Cold Snap in June, but cold snaps increased in spring. There has been a distinct shift of the temperature maximum toward end of July, with concomitant heat waves.

1.54
Annual temperature curve for Zurich: comparison of average values for two intervals. The lower part shows the deviation of the interval 1946–1985 from that of 1901–1945. The curve also documents a warming trend of 0.6 °C for the period 1946–1985.

1.55
This hourglass suggests time is running out for glaciers in our mid latitudes.

1.56
The Ried Glacier in the Swiss Alps. The tongue has retreated significantly since the 1860 maximum sculpted out the moraine walls of the glacier bed.

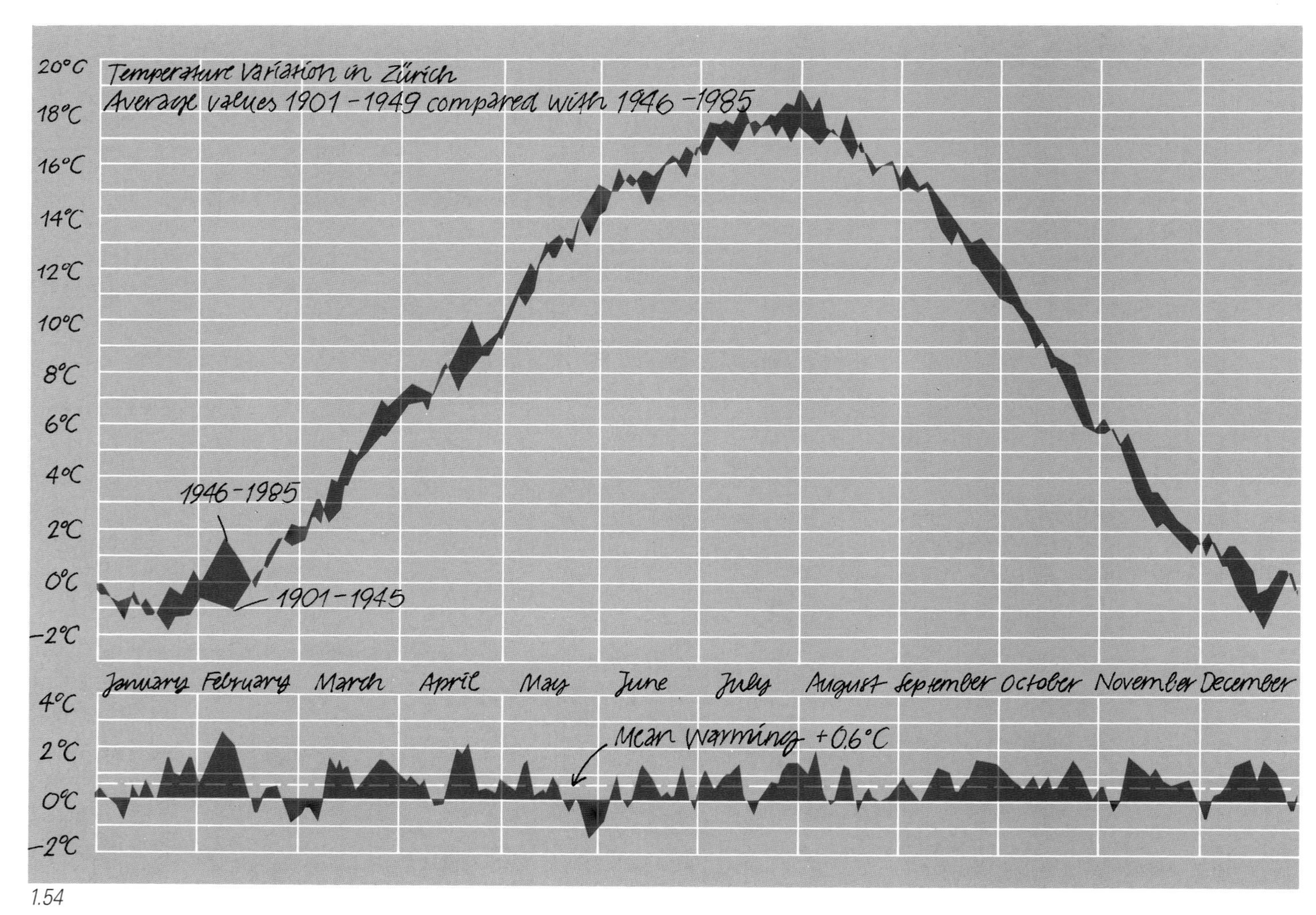

1.54

1.55

1.56

1 Precipitation and visibility

Temperature conditions are not the only controls on changes in glacier lengths. Amount and seasonal distribution of precipitation play a major role. The Meteorological Observatory in Neuchâtel archives one of the few internally consistent, long-term precipitation data sets, thanks to a continuity of traditional methods. These data also cover the 1901–1945 and 1946–1985 periods. No trends are obvious, even when seasons are compared individually. Fluctuations abound, but with irregular patterns. Must we conclude that the general retreat of glaciers, therefore, reflects a subtle increase in temperature?

Another change in our century has resulted in decreased visibility. Visibility is mainly determined by the relative concentration of particles in the atmosphere. Civilization has made its mark here by stirring up a lot of dust. Note the peak in Zurich during World War II due to increased coal and wood burning.

1.57
Curves for the amount of precipitation in Neuchâtel from 1901–1985 separated into seasons. The light blue background traces the smoothed curve which shows no distinct trends.

1.58
Plot showing the 100-year decreasing visibility of the Alps as seen from Zurich.

1.59
View toward the Swiss Alps across Berne. Nowadays, the panorama is only crystal clear during Föhn, and strong high-pressure situations.

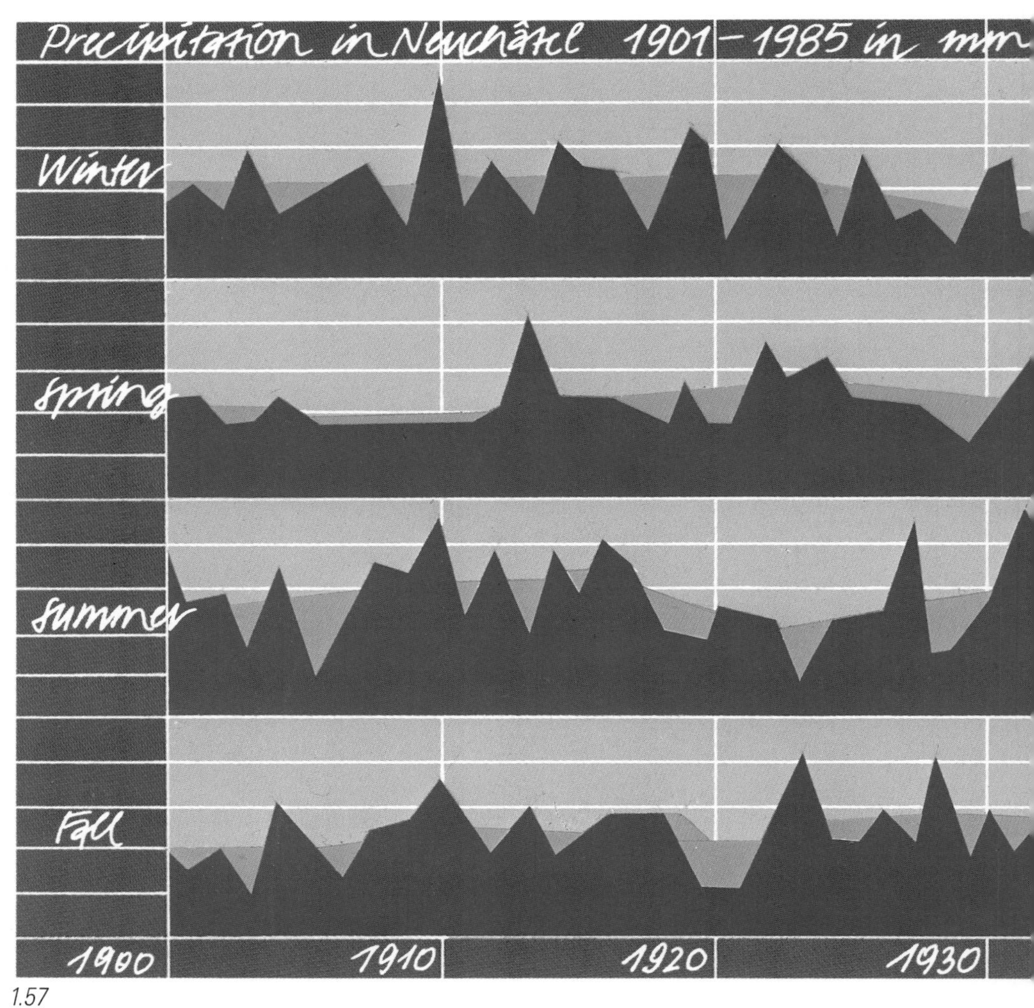

1.57

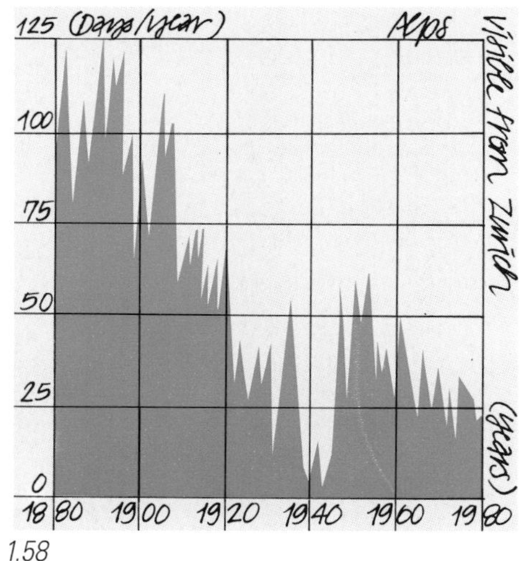

1.58

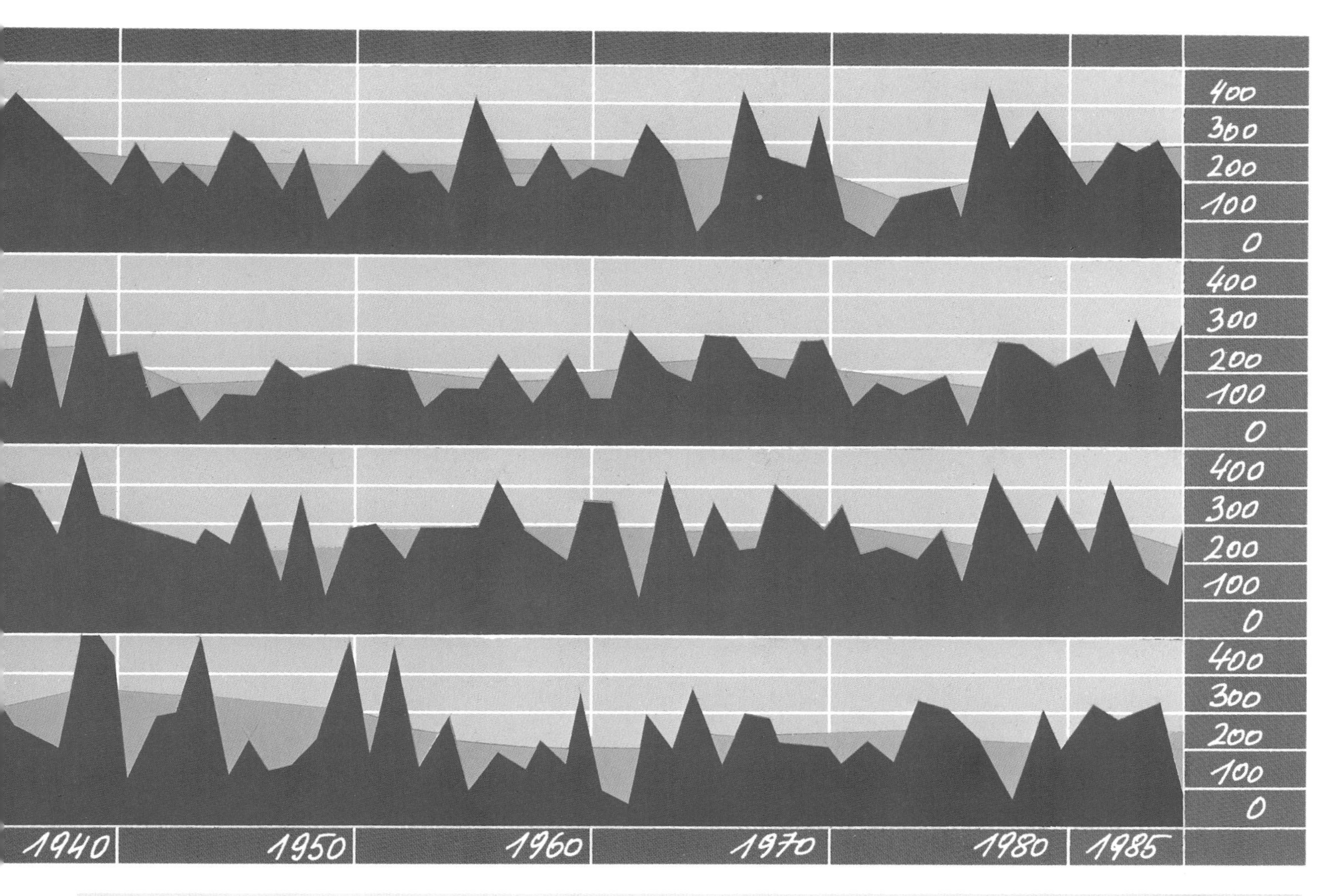
400
300
200
100
0
400
300
200
100
0
400
300
200
100
0
400
300
200
100
0
1940
1950
1960
1970
1980
1985

1.59

1 Weather and climate

Regional and local weather are manifestations of the global climate system. Over the past years, an increasing number of extreme weather events have been reported from all over the world. Floods, droughts, hurricanes and storm surges have brought considerable loss of life and economies.

There have also been some benefits but the losses seem to far outweigh the gains. Is this, because catastrophies sell better in the media than the normal or is it already a forecast of climate changes?

The man in the street and the policymaker are not interested in following the debate and criticism among scientists. They prefer facts and pat answers. What can the scientific community offer?The most developed tool for predicting future climate on a global scale is computer modeling using general circulation models. These models are based on the laws of physics, with additional, simplified descriptions, parameterisations, for smaller scale processes in the climate system. Such processes include cloud formation and dissipation, heat exchange between ocean and atmosphere, evapotranspiration, or sea-ice effects.

An accurate prediction of the magnitude and timing of temperature changes, precipitation, or extreme weather is limited: interacting parts of the climate system continuously change, thus leading to an unforeseen error propagation. What follows from present modeling experiments, therefore, are no pat answers, but rather a range of scenarios of what most likely will happen in the next decades on a global scale. Regional climate changes will differ from the global mean.

The global warming trend observed in instrumental data and predicted by models will lead to increased precipitation and evaporation of a few percent. The number of intense showers will also increase; and the number of days with temperatures at the high end of the temperature distribution will increase substantially. An unresolved question is whether the variability in weather patterns will increase. Presently, we have insufficient statistically comparable data to prove that this has already happened. From model experiments, there is also no clear evidence.

If warming shifts the position of depression tracks or anticyclones, variability and extremes of weather at a particular location would increase. The impact of the 1990 winter storms on the northwestern European coastline led to an increase in public discussions, but not to an improved understanding of their possible causes. Mid-latitude storms are driven by the equator-to-pole temperature contrast and this contrast will probably be weakened in a warmer world. Storms, therefore, might also weaken. But increased latent heat transport to higher levels in the atmosphere could also change storm tracks . . . at the moment a possible cause to think but not to model accurately on a global scale. Tropical storms, such as hurricanes, only develop at SSTs warmer than 25 °C. The extent of the world's ocean surface reaching such critical temperatures will increase in a warmer world. But will the critical temperature for tropical storm development also increase? At the moment, we do not know.

Better predictions of the future extent of snow cover and mid-latitude glaciers are possible. The conclusion that global temperature has been rising is strongly supported by the retreat of most mountain glaciers since the end of the 19th century. A 3 °C warming would melt most of the glaciers in the Alps.
The question still remains: How much confidence do we have in these predictions? The uncertainties do not alone arise from our imperfect knowledge of the climate system. The repsonse of climate, and also the validity of climate models, depend strongly on future rates of greenhouse gas emissions. Who is able to model our future behaviour?

What we know about climate

2

The planet Earth's climate is dynamic, constantly changing. The term, climate, itself is Greek, meaning slope. It derives from the latitudinal dependence of the height of the Sun over the horizon. We define climate in terms of atmospheric conditions averaged over a long time period, which can be measured in years or even millions of years. Our scope is set by the time scale with which we choose to observe the events on the surface of the Earth. In this book, we limit our view to the last few million years, which are characterized by major oscillations between cold and warm periods, with emphasis on the period of the last glacial cycle. We must ask ourselves, what can we learn from the lessons of the past about our own future?

2 The climate system

We now know that climate is controlled by complex feedbacks and interactions among the individual components of the climate system. Simplified, these include solar activity, the atmosphere, the hydrological cycle and the biosphere. The temperature of the Earth's surface depends on radiation from the Sun, plus the Earth's momentary distance from the Sun, plus the transparency of the atmosphere for radiation, plus the reflectivity of the Earth's surface. Temperature gradients are created by decreases in solar radiation from the equator to the poles. Snow and ice increases lead to further cooling. Dark areas of the earth's surface are warmed. The hydrological cycle is driven by differences in evaporation, precipitation, wind and circulation in the seas. Ultimately, hydrology determines the location of vegetational belts and cold or hot deserts. The elements of the climate system are so sensitive to their mutual interactions or disturbances from outside, that we cannot totally comprehend or predict the effects precisely. This we call open sytem behavior. The individual processes are described by strict physical laws. Today, and tomorrow, we are simply confronted with a climatic situation to which we can merely adapt.
We can learn about the climate conditions in the past thanks to Mother Nature, who archives her own experiments in subtle ways. With the help of state-of-the-art physical, chemical, and biological methods, these secrets are decoded step by step.

2.1
The Sun is the dominant motor in the climate system. The position of the Earth with respect to the Sun is a constantly changing parameter, which also forces the climate. Sunset in Finnland.

2.2
The Atmosphere, a thin, invisible veil, protects the Earth's inhabitants, making life possible. The composition of atmospheric gases controls the temperature budget by the absorption of long-wave infrared re-radiation from the Earth's surface. Thunderheads in Kenya.

2.1

2.2

2.3

2.3
The Oceans and the Hydrologic Cycle. The oceans store 97% of the water. Solar energy drives the hydrologic cycle. Evaporation, vapor transport, and precipitation determine the climate zones and thus define the conditions for life on our planet. McMurdo Sound, Antarctica

2.4
The Biosphere is the special aspect of our Earth among the planets. The Earth's atmosphere and its elemental cycles are regulated by the biosphere. The climate is strongly modulated by the biosphere via its control on the carbon and methane cycles. Cells of a walnut tree-ring.

2.5
Archives store the changing fate of our Earth and its climate like a scrambled, tape-recorded message. Examples of natural archives are polar ice caps, ocean and lake sediments, tree rings, peat bogs, fossil soil profiles, and glaciers and moraines. Tree-ring textures from an alpine hut.

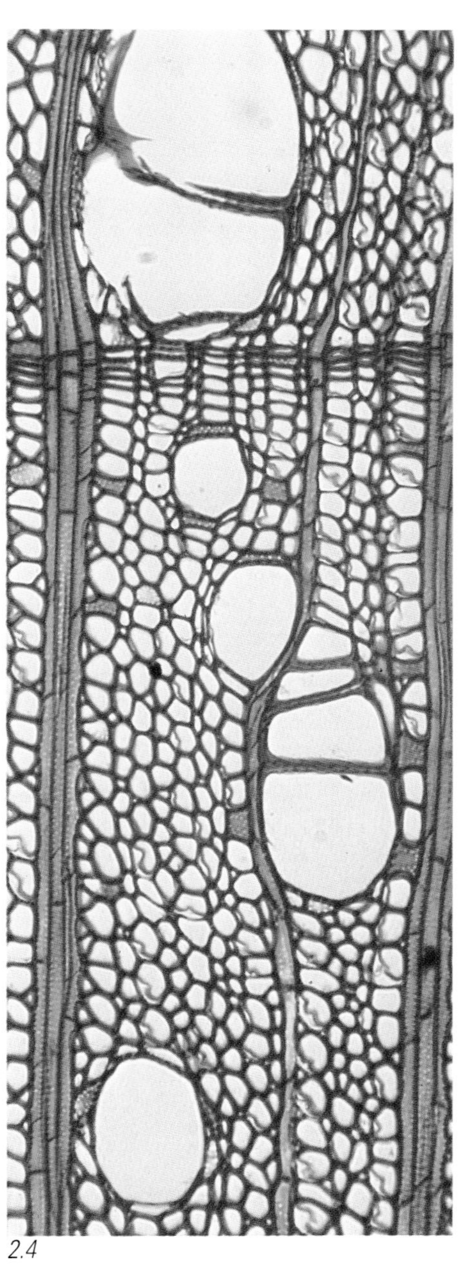

2.4

2.5

2

The Sun

The Sun is the energy source and motor of the climate system. To what extent do solar events directly affect our climate? We cannot yet say with certainty. Scientists believe that solar radiation intensity follows a long-term cyclic pattern. Sunspots are dark areas on the surface of the Sun that have been observed for centuries. Their numbers vary according to an 11-year rhythm. They are related to periods of increased solar activity and short magnetic storm events on the Sun's surface. The sunspots reduce the solar radiation by a minute fraction, but extended satellite measurements today show that the total solar energy reaching the Earth during these events increases. Sunspot records thus monitor solar activity. The *solar constant* is an expression for the emitted energy. This value is known to vary only tenths of a percent. This affects the Earth's temperature by less than 1 °C.

2.6
The Sun's intensity is measured by a pyrheliometer, which is calibrated at the World Radiation Center in Davos, Switzerland.

2.7
Graph of Sunspot abundance 1700–1985.

2.8
In 1848, the Swiss astronomer, Johann Rudolf Wolf introduced the Sunspot Index as a consistent method of monitoring Sunspots.

2.9
A computer simulation of changes in solar radiation caused by sunspots. The asterisk shows the value corresponding to the Sun's image for June 11, 1969.

2.10
Tremendous amounts of superheated hydrogen gas are thrown into space during eruptions on the Sun.

2.6

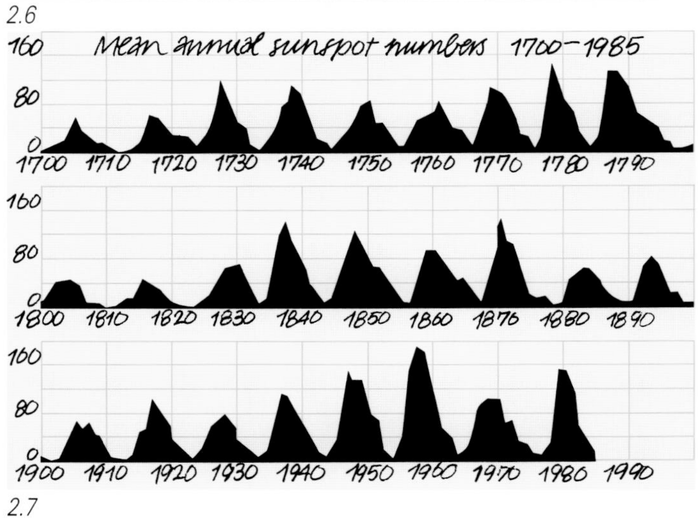

2.7

2.8

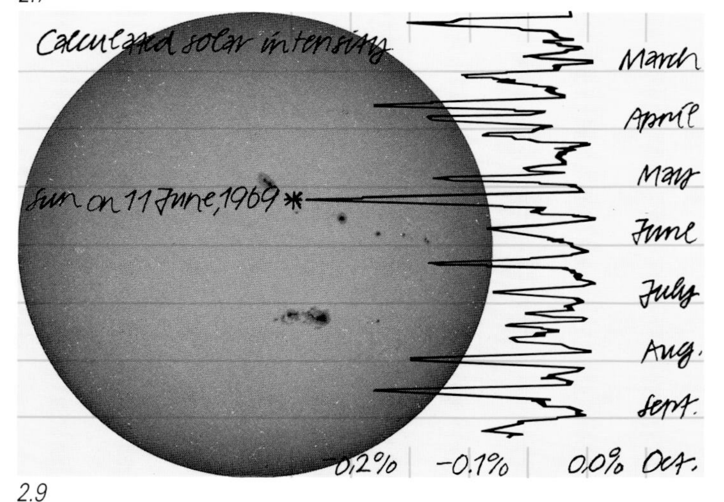

2.9

2.10

2 The Earth and Sun

Everyone notices that days get longer or shorter and that summer in the Northern Hemisphere means winter for the inhabitants of the Southern Hemisphere. The reasons are described by the *orbital parameters: Eccentricity* is a small variation in the shape of the orbit of the Earth around the Sun due to interactions with the orbits and gravitational pull of the other planets. *Obliquity* is a slight change in the tilt of the axis of rotation of the Earth with respect to the orbital plane. Finally, like a top, the spinning Earth *precesses,* with its axis moving slowly around a circular path. Each fluctuation has its own frequency, which results in more or less solar input, *insolation,* at a given point on the Earth's surface. The Yugoslav mathematician, Milutin Milankovich, was the first to calculate the insolation curves for various latitudes based on the orbital laws. These curves help explain repeated switches from ice ages to warm intervals which are recorded in natural archives worldwide.

2.11
The degree of orbital *eccentricity* changes the relative Earth-to-Sun distance on a 100 000-year cycle.
2.12
The *obliquity* of the Earth's axis with respect to the orbital plane tilts between 21.5° and 24.5° with a cycle of 41 000 years.
2.13
The Earth's axis *precesses* in cycles of 19 000 to 21 000 years.
2.14
Calculated past and future changes of eccentricity, precession and obliquity.
Below. The curve calculated by Milankovich for insolation in summer shows that during warmer times, 65 °N would receive the insolation of 60 °N today. During ice ages – blue filled area – 65 °N would receive as little insolation as 80 °N today.
2.15
Path of the midsummer night's Sun in Lapland. The tilt of Earth's axis provides visibility of the Sun during 24 hours.

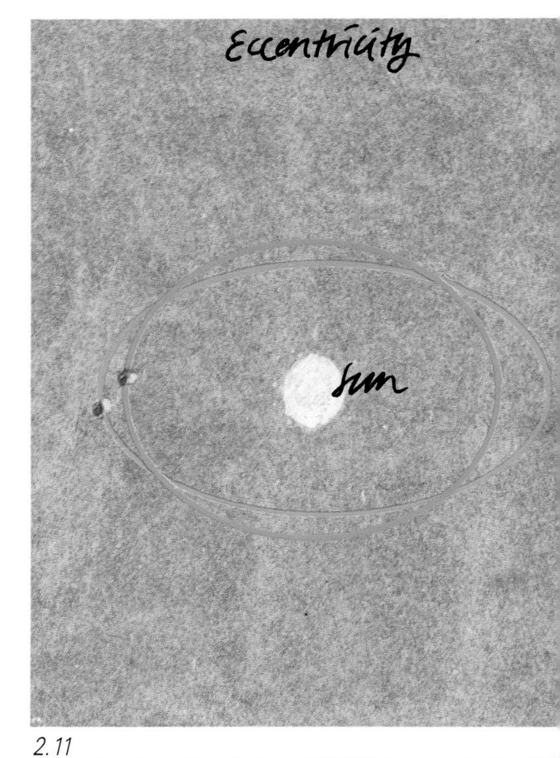

2.11

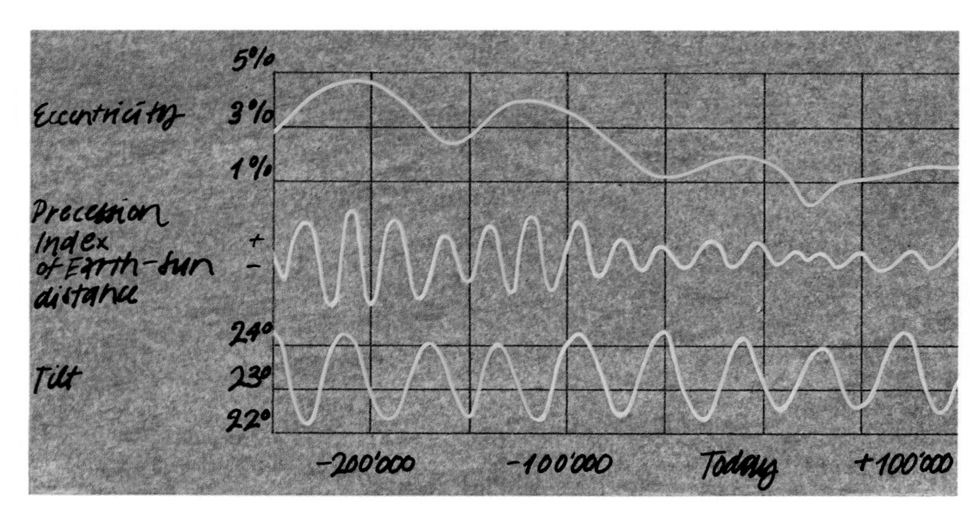

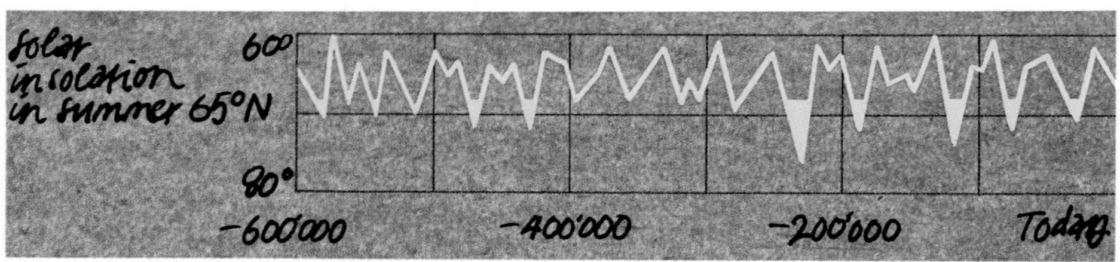

2.14

2.15

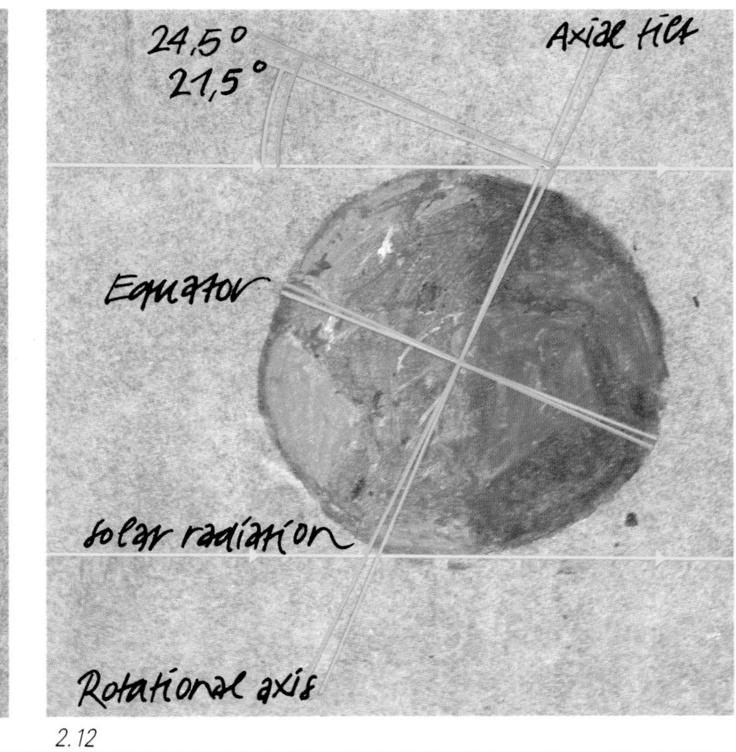
24,5°
21,5°
Axial tilt
Equator
solar radiation
Rotational axis
2.12

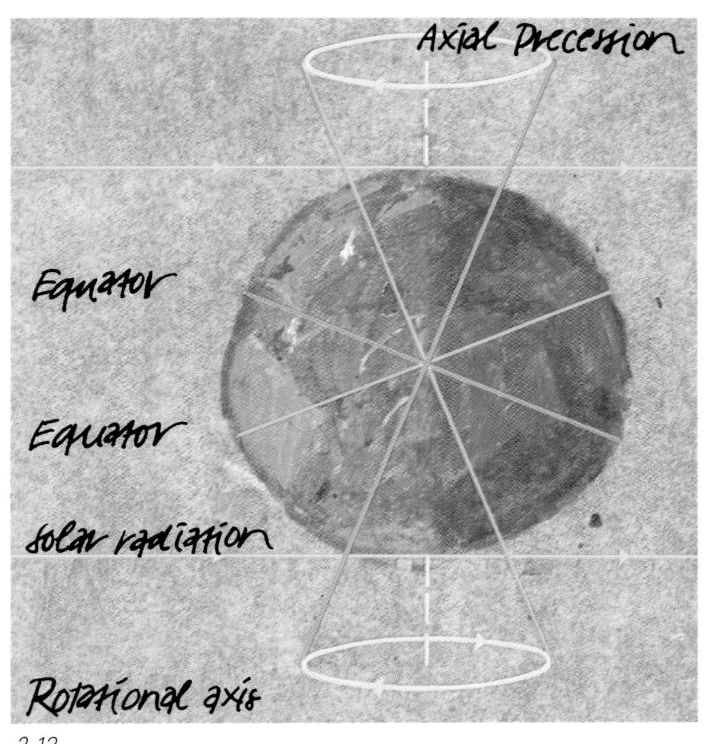
Axial Precession
Equator
Equator
solar radiation
Rotational axis
2.13

2 The atmosphere

During past centuries, the atmosphere was literally invisible to the scientific community. This has changed radically since we have recognized its significance for weather and climate – and the breathtaking pace of industrialization in some areas. The atmosphere shields us from bombardment by cosmic particles and the deadly ultra-violet rays of the Sun. The Earth not only captures and holds the Sun's energy; it also returns a portion to space. How much depends on the composition of the atmosphere. Water vapor, dust and, increasingly, greenhouse gases such as carbon dioxide, methane and ozone control the input-output radiation balance. We live, therefore, in a delicately regulated greenhouse environment. Our blue planet has been spared the harshness of the moon, without any atmospheric moderation, or the boiling hell of Venus with its dense atmosphere, rich in carbon dioxide.

2.16
Sunrise behind the Earth's horizon backlights the thin veil of atmosphere, as seen from *Apollo 12* returning from the Moon, November 1969.

2.17
Northern Lights, the *Aurora Borealis,* above Kirunà, Sweden. Ionized particles from the Sun excite atoms in the upper atmosphere which then release energy as light. The atmosphere is alive.

2.18
The scheme of the atmosphere projected onto a backdrop of the Swiss Alps. Weather mainly involves the troposphere. The temperature profile through the atmosphere decreases uniformly up to the stratosphere, then increases in the ozone layer due to absorption of ultraviolet rays.

2.16

2.17

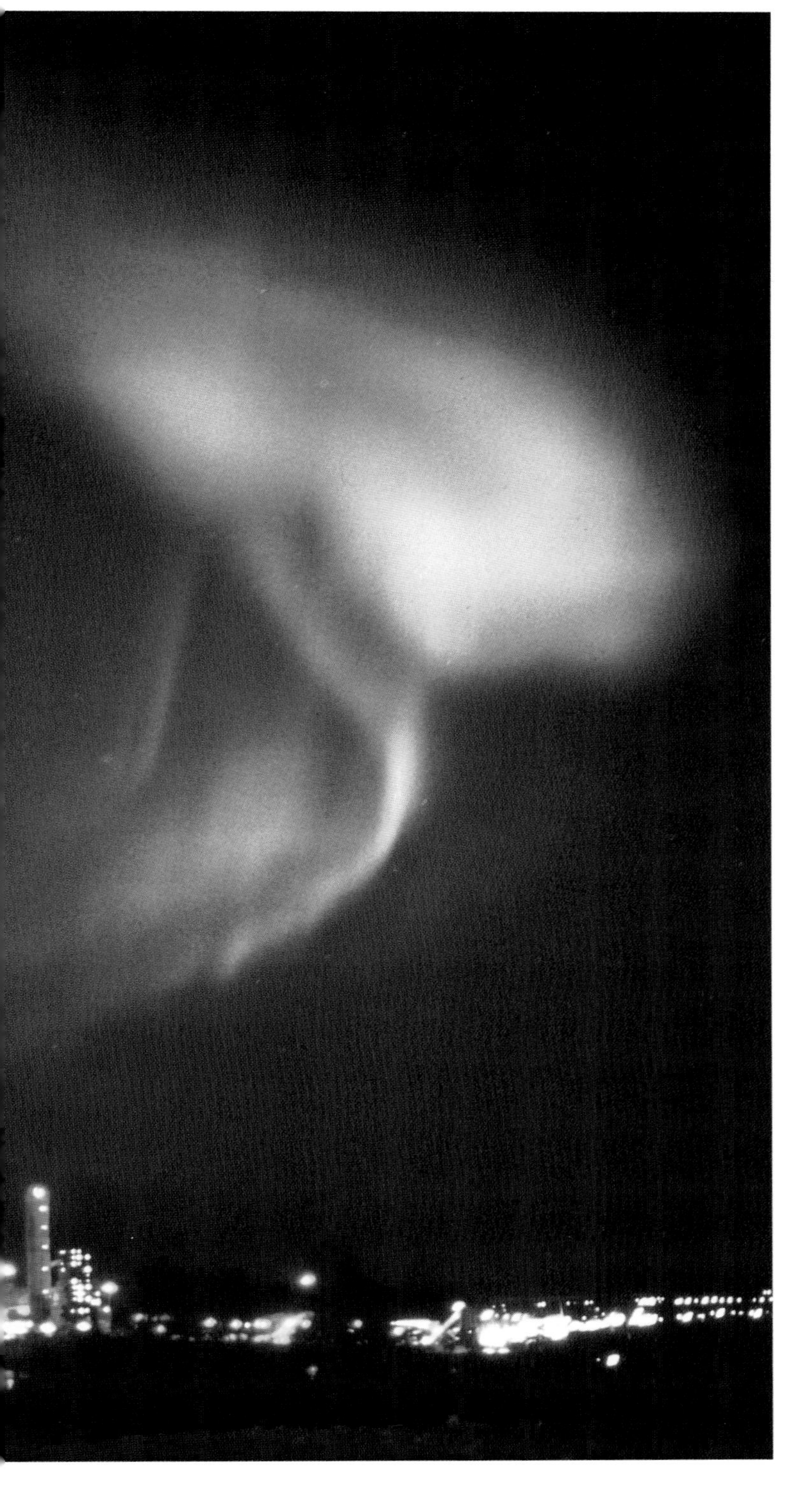

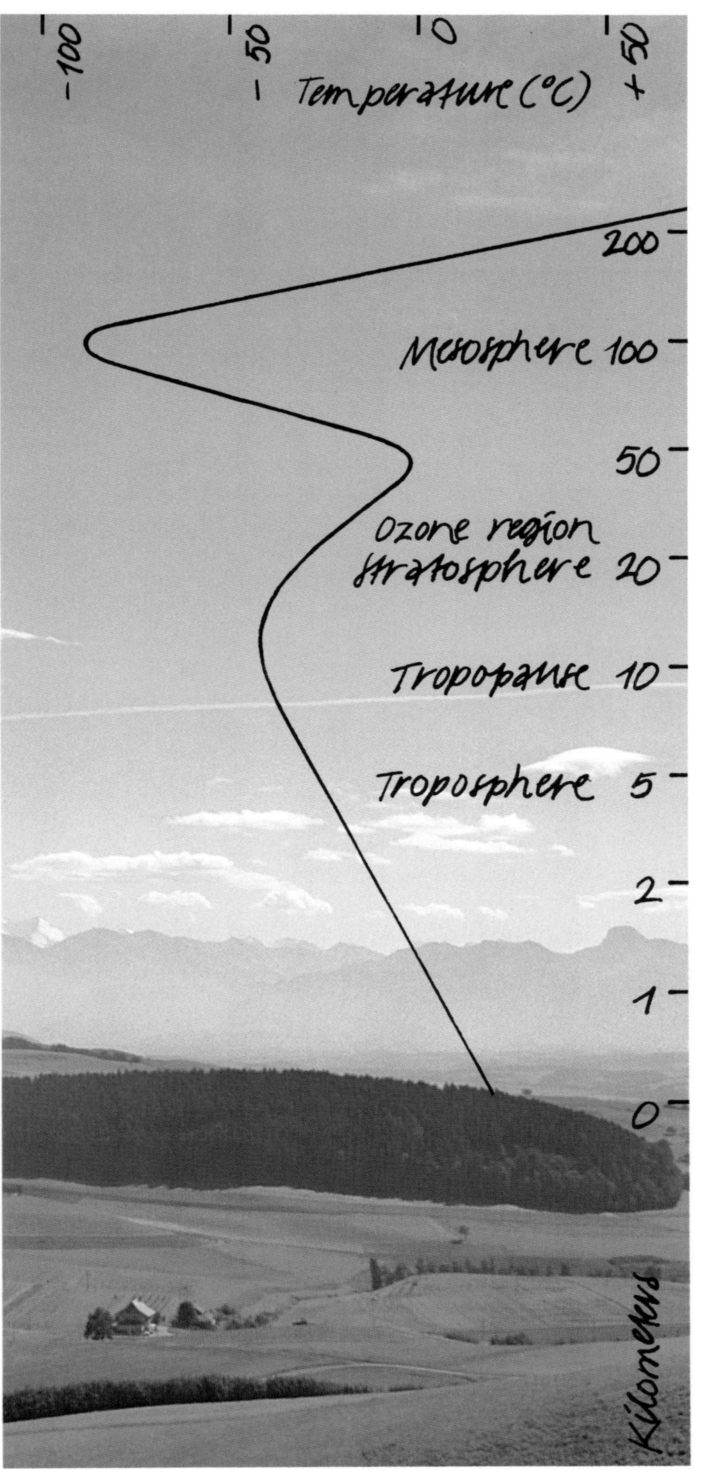
-100
-50
0
+50
Temperature (°C)
200
Mesosphere 100
50
Ozone region
Stratosphere 20
Tropopause 10
Troposphere 5
2
1
0
Kilometers

2.18

2 The energy budget of the Earth

We have learned how the temperature of the surface of the Earth is determined by the amount of solar radiation penetrating the atmosphere. Heating from the Earth's interior, or that produced by man is insignificant. Most of the incoming short-wave radiation from the Sun penetrates to the Earth's surface. The remainder is either trapped and absorbed in the atmosphere, or reflected by cloud surfaces and dust. The Earth returns its energy in the longer-wave, infrared spectrum, because its surface is cooler than the Sun's. Water vapor, carbon dioxide, methane, and other trace gases make the atmosphere less transparent to the outgoing long-wave radiation; thus most of this is radiated back to the Earth. This warms near-surface air masses much like air in a greenhouse under protective glass. Without this natural greenhouse, we would chill down to −18 °C, rather than remaining at today's comfortable +15 °C. Rare, large volcanic eruptions spew millions of tons of ash and gases into the stratosphere, leading to natural changes in the Earth's energy budget. Humans now excercise an equal impact on the natural regulatory mechanisms due to their additional production of greenhouse gases, particularly carbon dioxide, the chlorfluorocarbons, nitrous oxide, and methane.

2.19
The Earth's energy budget expressed in percent of solar radiation reaching the top of the atmosphere. Incoming short-wave radiation from the Sun is absorbed by the Earth's surface and re-emitted as long-wave, infrared radiation. Because the atmosphere is not very transparent for infrared, most radiation is absorbed by the atmosphere and returned back to the Earth's surface. Latent heat transferred by evaporation and precipitation is also absorbed by the atmosphere and adds further drive to the weather machine.

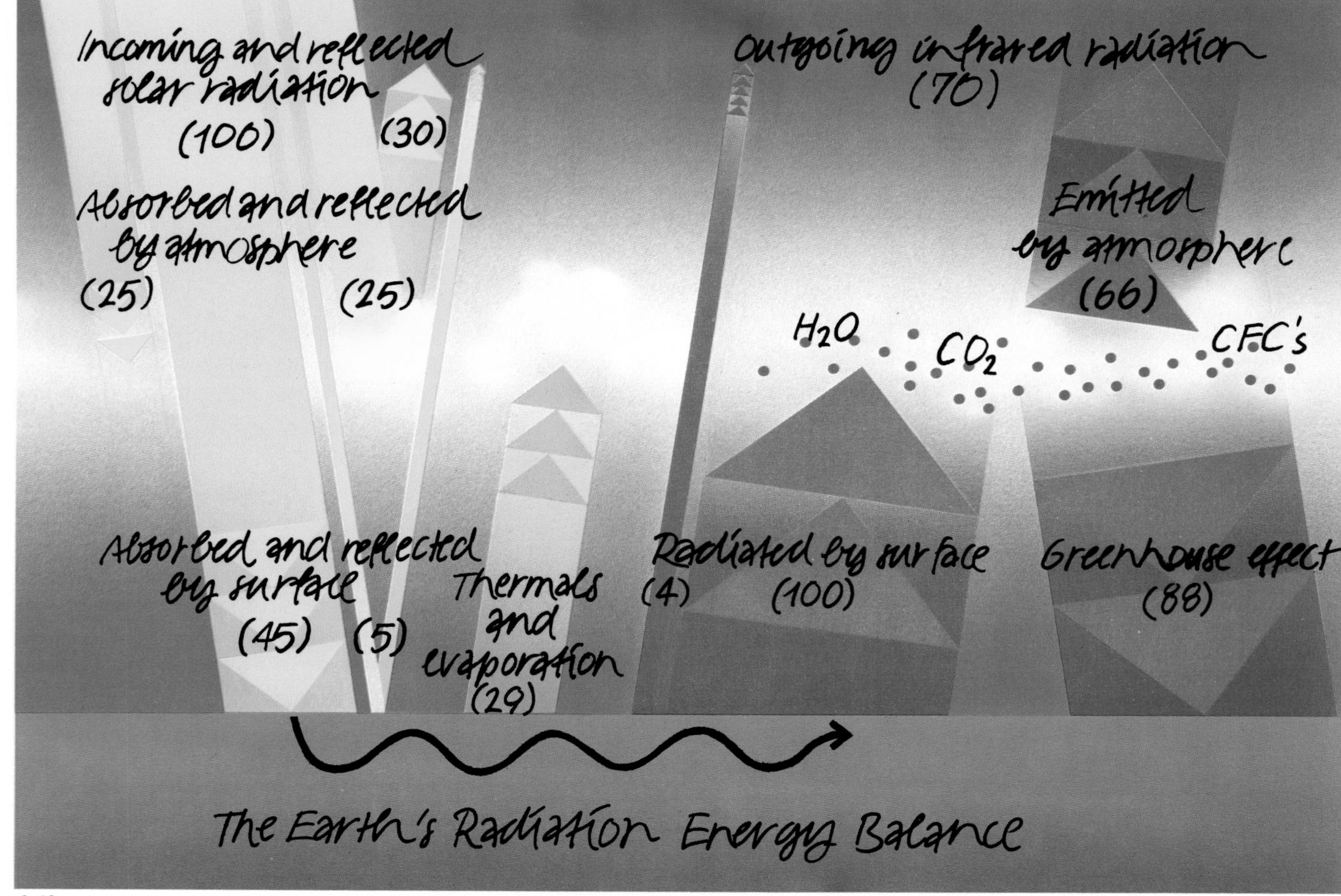

2.19

2.20
Sunset in February 1983 in a Swiss valley. This seasonally unusual intensive color was caused by the atmospheric dust from the eruption of El Chichon Volcano, Mexico.
2.21
The effect of volcanic eruptions on the incoming solar energy measured in a specific atmospheric layer.
2.22
Eruption of Vesuvius in 1822.

2.20

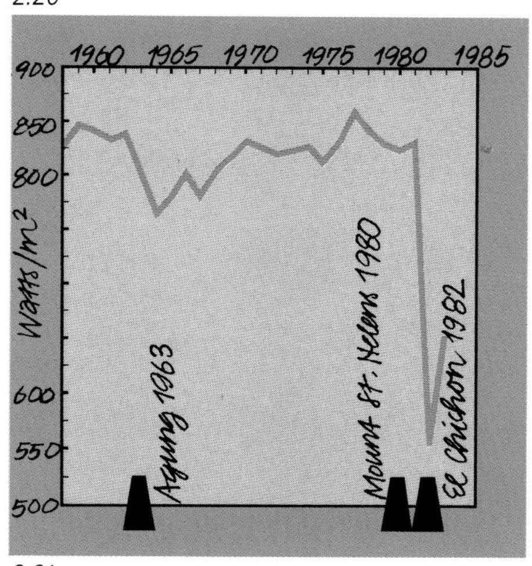

2.21

2.22

2 Oceans and the global water cycle

The ocean and atmosphere are intimately coupled with the global water cycle. Cooling or warming of the sea surface directly controls the characteristics of overlying air masses; here lies the cauldron for weather. The ocean has a positive radiation balance: more radiation is absorbed than emitted. The excess heat is dissipated by evaporation or direct conductive heating of the atmosphere.

2.23
Seventy-one percent of the Earth's surface is covered by the oceans, with a non-uniform distribution of landmasses.

2.24
Water spouts form when gusts of polar air brush across warm ocean water, and beautifully portray the coupling of ocean and atmosphere.

2.25
Model of the global water cycle. The thickness of the arrows is proportional to water fluxes, in units of 1000 km^3 of water per year.

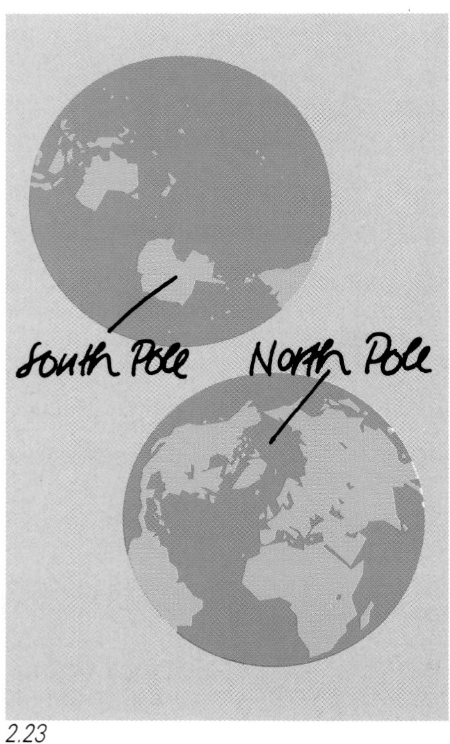

2.23

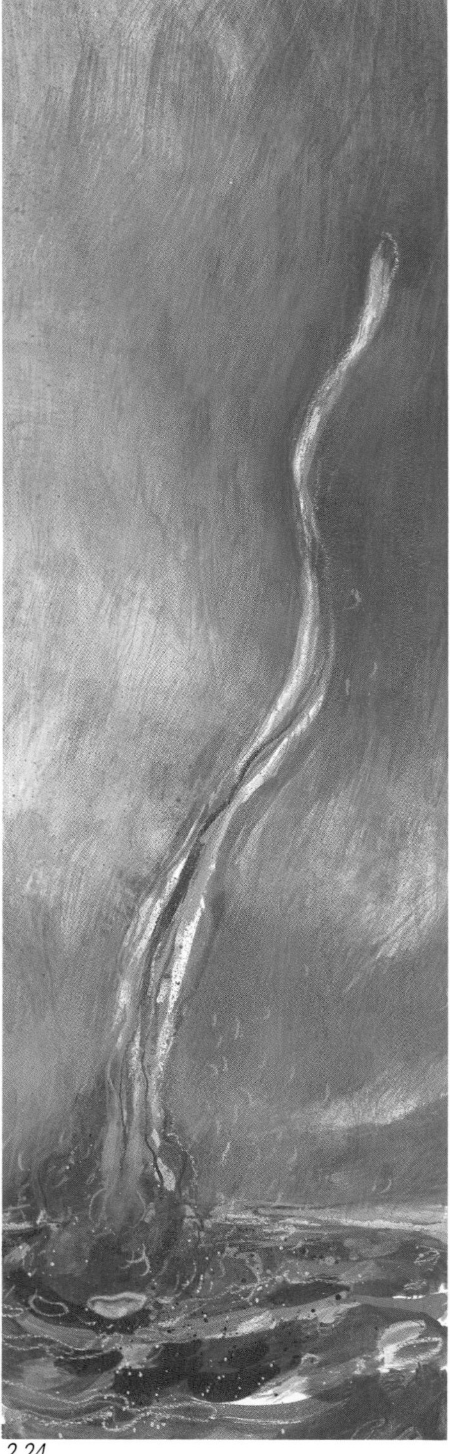

2.24

2.25

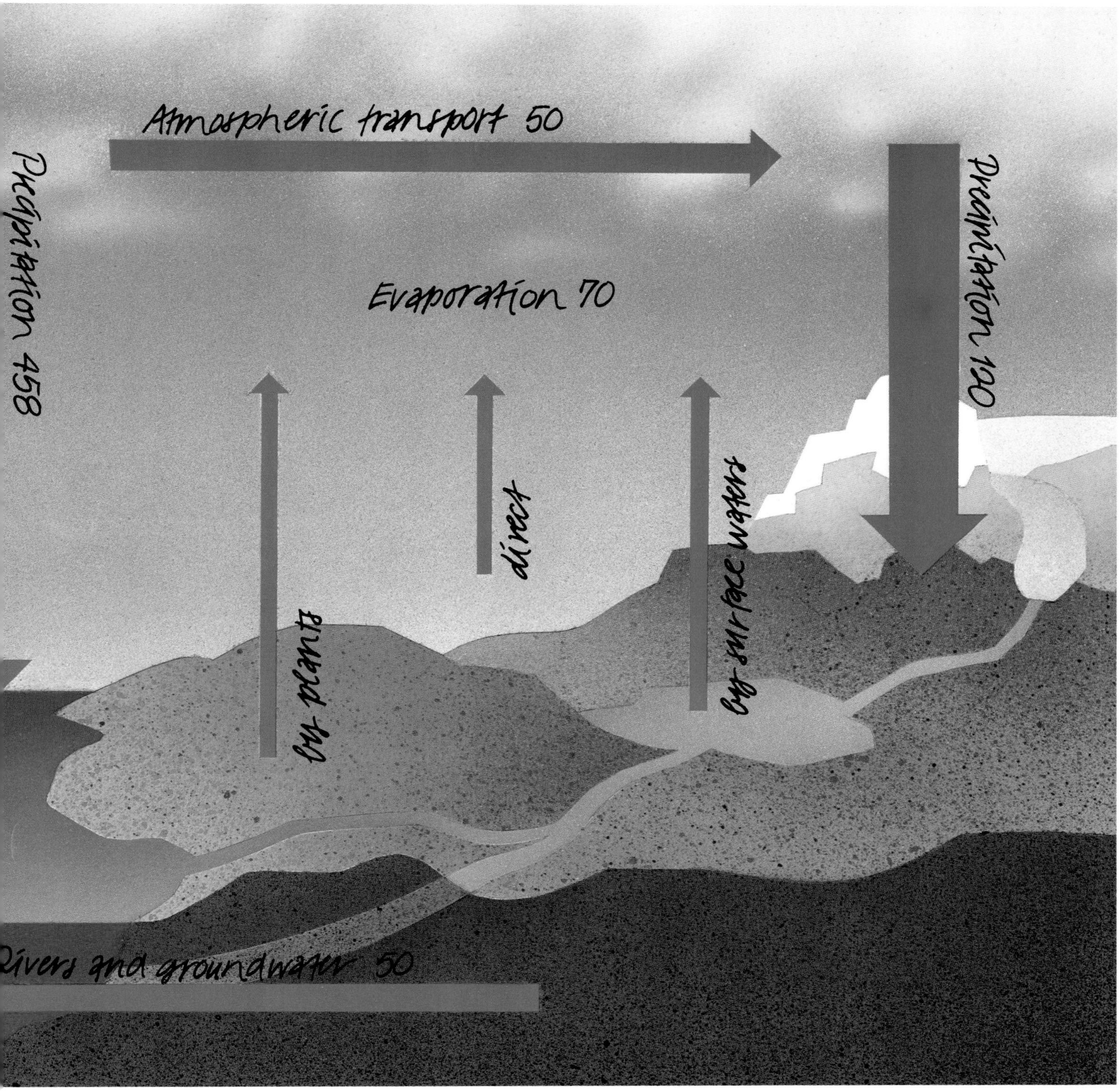

Atmospheric transport 50

Precipitation 458

Precipitation 120

Evaporation 70

by plants

direct

by surface waters

Rivers and groundwater 50

2

If we examine that part of the global water cycle driven by evaporation, we realize that the transfer of water, ocean – atmosphere – ocean, is more significant than the detour across the continents. Changes in amounts of water stored in polar ice caps or glaciers, on the other hand, play a determinant role for large-scale climate patterns. Concomitant rises in sea level caused by warming are among the most threatening effects for society. Potential sea-level rises provoke media images of The Flood and Apocalypse.

Sea-surface temperature – SST – is dependent on latitude; the thermal excesses at low latitudes are balanced by thermal losses toward the poles. Wind, temperature and salinity drive ocean circulation, which transfers immense thermal energy. The regions with SST greater than 10 °C can be considered thermal sources. Today, the intersection of the 10 °C isotherm is found at approximately 55 °N and 45 °S. These are called the northern and southern Polar Fronts. Turbulent oceanic eddy currents are common along the polar fronts and, together with deep-water circulation, they play an important role in energy transfer within the oceans.

2.26
A temperature cross section through the western Atlantic shows that a relatively small water mass forms a zone of thermal surplus.

2.27
Offshore winds drive water masses toward the open sea and cause the upwelling of cooler water from deeper layers.

2.28
Schematic picture of a deep-reaching eddy spiral along the Polar Front. Measurements from current meters hung from buoys show that eddy-driven currents can reverse direction with depth.

2.29
Average heat transport in the oceans. Winds and currents drive the exchange.

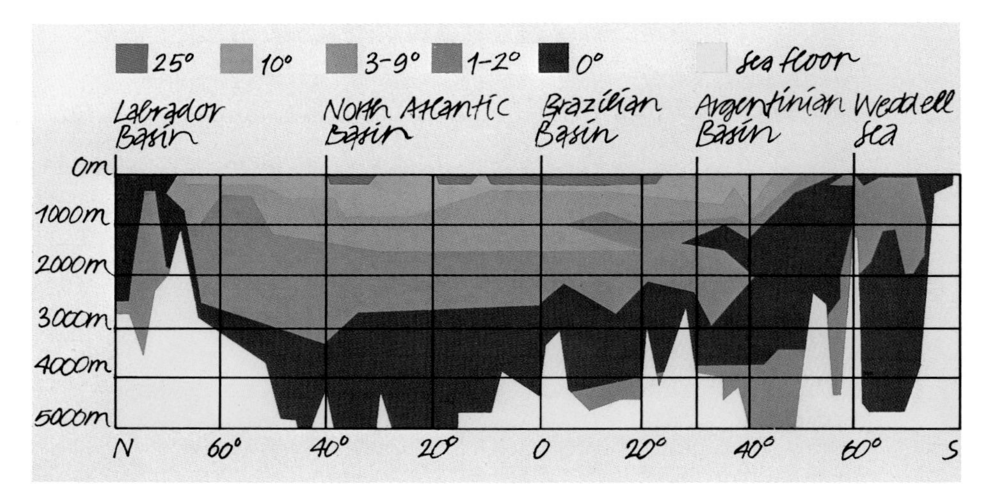

2.26

2.27

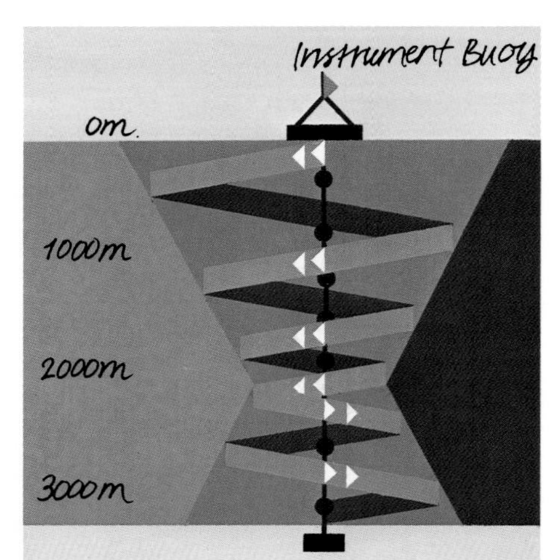

2.28

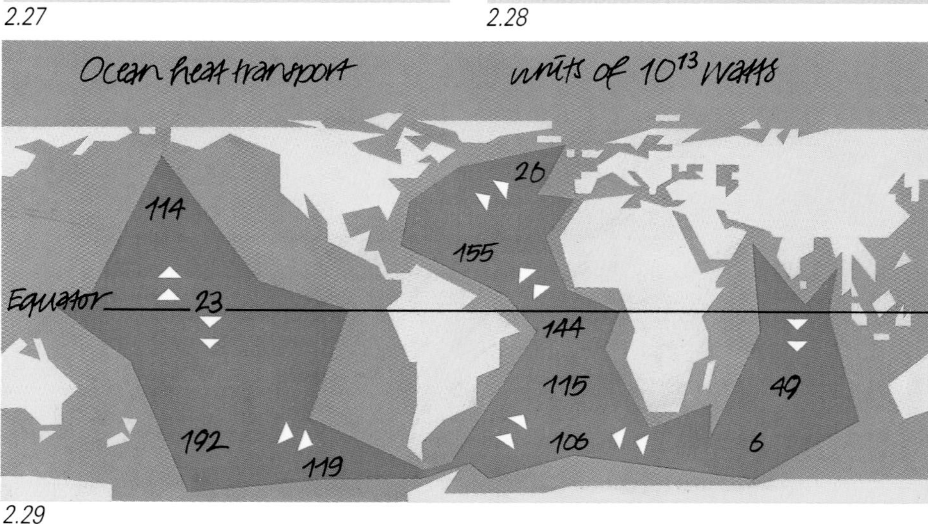

2.29

2.30
Satellite image of the Gulf Stream. The false color scale shows temperatures from 28°–24 °C in red and orange, 23°–17 °C in yellow and green, 16°–10 °C in blue, and 9°–2 °C in purple. The Gulf Stream flows like a giant, warm river from the Gulf of Mexico along Florida (1), leaves the coastline at Cape Hatteras (2) and loses itself in warm (3) and cold (4) eddy meanders as it progresses northeastward. By the time the Gulf Stream reaches the middle of the Atlantic, its surface has cooled sufficiently to make it invisible to satellites. As an important warm-water heater for Europe, the Gulf Stream can steer climate fluctuations, producing either very warm, nearly snow-free winters, or extremely cold winters, so cold that even the Rhine River freezes. We know too little about the further fate of the Gulf Stream in comparison with its importance. We will later show what happens to Europe if this natural warm-water tap is turned off.

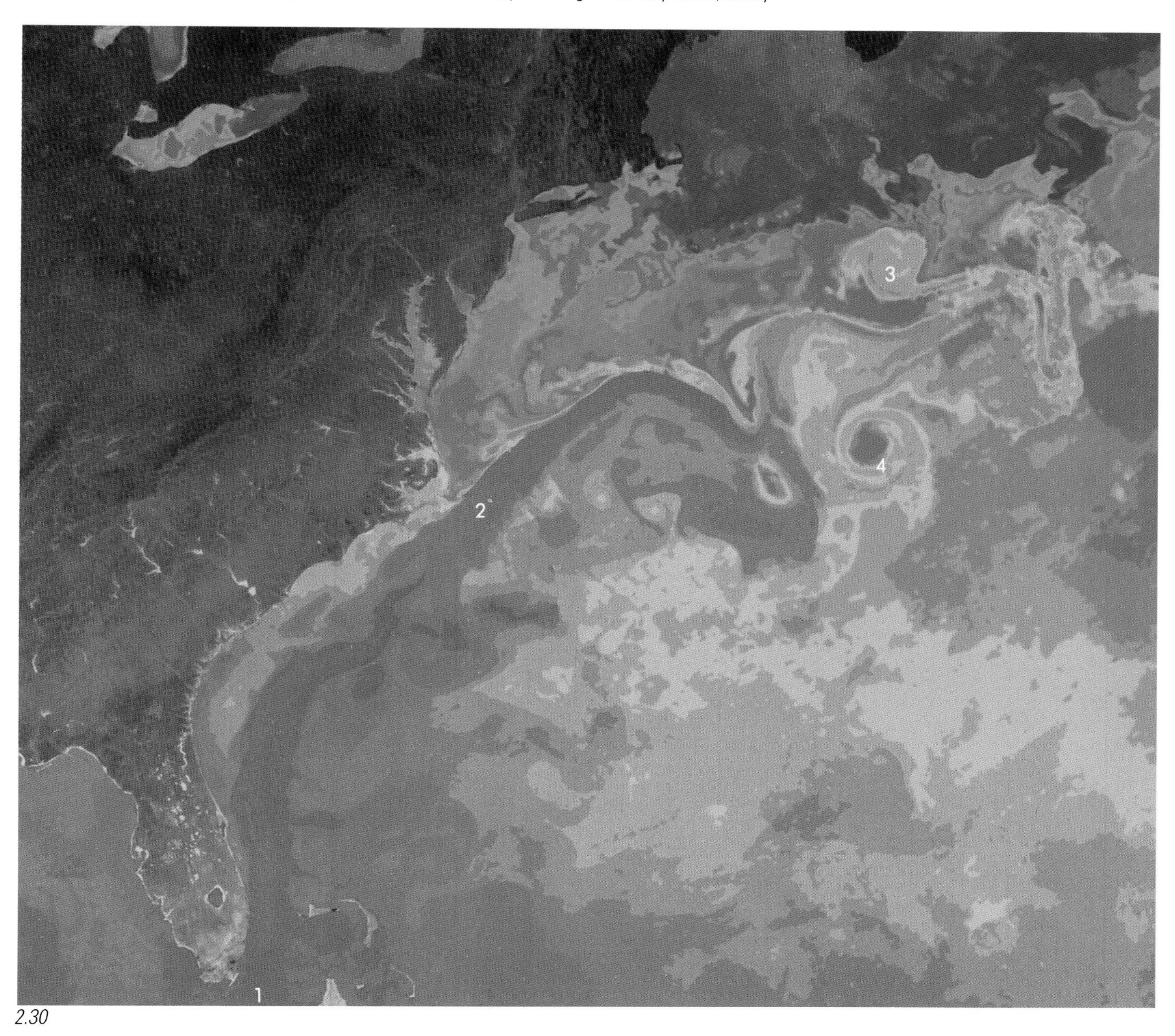

2.30

2 The biosphere

The Austrian geologist Eduard Suess introduced the term *biosphere* to science over one hundred years ago. The biosphere encompasses the entire part of the Earth that supports life. Reduced to its essence, the role of plants in the biosphere is to convert solar energy into organic matter via photosynthesis of carbon dioxide and water with concomitant release of oxygen. The phytoplankton in the oceans is responsible for producing about 70% of our oxygen, and serves as the major transfer agent for atmospheric carbon dioxide. Thus the living and dying of marine biota acts like a giant carbon-dioxide pump, which has kept the atmospheric CO_2 concentration remarkably constant over the last million years. The biospheric control on the greenhous gases CO_2 and methane appears to couple the Northern and Southern Hemispheres during climate changes.

2.31
The organic carbon cycle begins with photosynthetic production of plant tissue from atmospheric CO_2 and water. Plants form the basis for diverse food chains. Microorganisms degrade dead matter; carbon is then oxidized and released again into the atmosphere by soil respiration. A portion of the dead organic matter is sedimented, buried and, in the course of past hundreds of millions of years, has formed our fossil fuels: oil, gas and coal.

2.32
Tropical forest on the flanks of Kilimanjaro. Forests fix CO_2. Deforestation is an expensive display of the lack of human understanding for natural cycles.

2.33
That which was formed and stored in hundreds of millions of years, we burn in just two hundred...a million times faster! How will the global biosphere pump react to this pulse? Can it counterbalance the human interference in the climate system?

2.34
An outcrop of carbon deposits. About 6000 years ago a pine tree stump was buried in a moor in Scotland and archived for study today. Tree rings tell us about climate history.

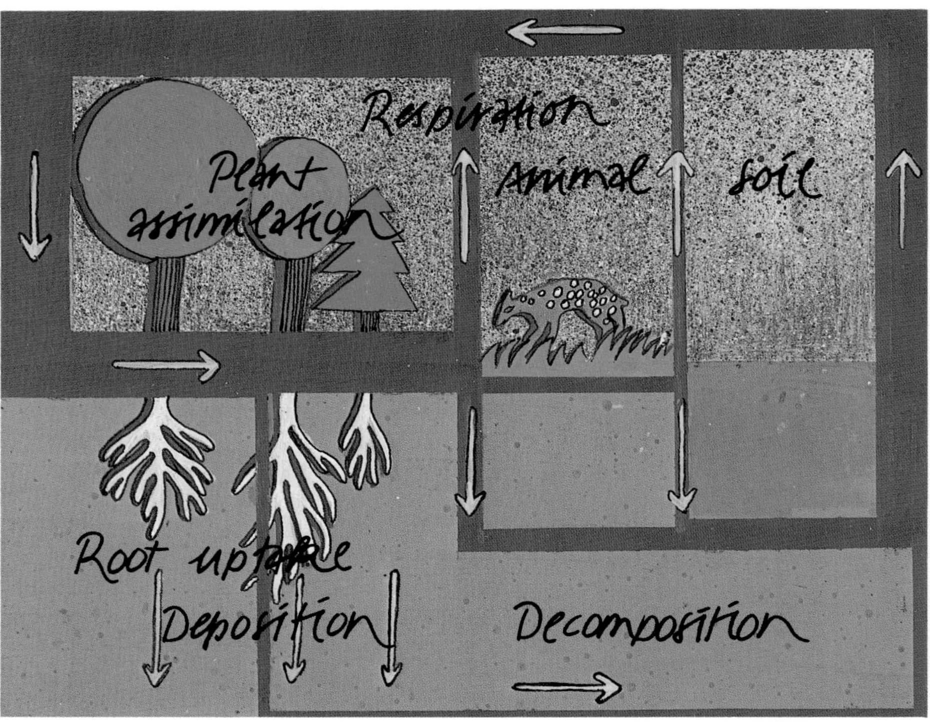

2.31

2.32

2.33

2.34

2

The archives

How can we learn about climate change that occurred prior to the age of daily instrumental weather monitoring? A first step is to examine the records of human history. Science is also learning to read what we call *proxy data.* These are bits of climate information encoded on the tape recordings in various natural archives. Tree rings, glacier moraines, polar ice, fossil soils, and lake and deep-sea sediments register different kinds of signals from Nature's experiments, including evidence of climate change. The job of the scientist is to understand these experiments and to convert the proxy data into past temperature or precipitation values. This is a difficult task. In order to resolve the duration and rate of change of past climatic events, we must precisely match the age of different archive slices. In reality, the scientific interpretation is akin to listening to only one noisy channel of a stereo tape recording which has been scratched by a cat's paw. The more we learn from the lessons of the past, the better our chances of planning a future in harmony with natural systems.

Sediments

Sediment layers contain microfossils, pollen and geochemical signatures that can be used to reconstruct environmental conditions for millions of past years. In many lakes and in some ocean basins, the seasonal rain of particles from pollen, algae, calcium carbonate, or dust lie undisturbed. These rhythmic sequences are called *varves;* they contain a record, similar to tree rings, of information on the responses of ecosystems to changing environments.

2.35
Coring of deep-sea sediments with a drill ship. The ship maintains position over a drill site with the help of sideward thrusters controlled by computer monitoring of sonar buoys on the seafloor. The R/V *Glomar Challenger* recovered thousands of cores from the world's ocean from 1968 onward and has now been replaced by the Ocean Drilling Program ship, R/V *Joides Resolution.*

2.36
A sediment core from Lake Van, Turkey displays vividly laminated couplets, *varves,* formed by an annual sedimentation cycle.

2.35

2.36

Polar ice and glaciers

Cold glaciers – those with an average ice temperature much below zero centigrade – and polar ice caps store precipitation in a giant deep freeze, with records going back to about 300 000 years ago. Falling snow flakes harvest particles and gases from the atmosphere and pile layer upon layer of information. As ice forms under these increasing loads, the direct history of the Earth's atmosphere becomes captured and imprisoned in ice crystals and bubbles. Even seasonal information is preserved for hundreds of years across endless expanses of Antarctica and Greenland. In contrast, the coldest summit regions of mid-latitude glaciers are often subjected to strong winds, which remove part of the snow cover, making these archives more difficult for scientists to interpret.

2.37
A Swiss team cores a shallow site on the Greenland ice cap.

2.38
A snow vane over the Matterhorn as seen from a drill site at 4500 m on the Monte Rosa Massif. The 124 m of ice, cored to bedrock, covers the last 1000 years. The altitude and alpine conditions limit work in such regions.

2.39
Sahara dust and ice layers outcropping below the Monte Rosa drill site.

2.38

2.37

2.39

2 Moraines

The position of a glacier snout fluctuates with climate as an obvious expression of more or less precipitation and higher or lower temperatures. Advancing glaciers override soils and forests, bulldoze moraine walls and gouge the bedrock with ice-held stones. During the last 10 000 years, European glaciers have rarely extended beyond the maximum limits marked by the 1850 moraines. Organic material dug from moraines is dated to provide a time framework for fluctuations. Moving mountain glaciers write a local climate history on their surrounding landscape. Matching these fluctuations to the information in polar ice will add significant pieces to the global climate puzzle.

2.40
Over-ridden tree trunks imbedded in a moraine near Arolla, Switzerland.

2.41
Air photo of moraine ridges near Arolla.

2.40

2.41

2.42
The 1850 advance of the Great Aletsch Glacier as it overrides a forest.
2.43
Glacier striations on bedrock near Cuzco, Peru. Nearby, climate history is being deciphered from the Quelccaya ice cap.
2.44
Sketch of moraines between the 1850 maximum and the position of the glacier front today. Most of the intermoraine regions have been repopulated with a pioneer vegetation. Such localities provide open-air classrooms for studying vegetation dynamics.

2.42

2.43

2.44

2

2.45

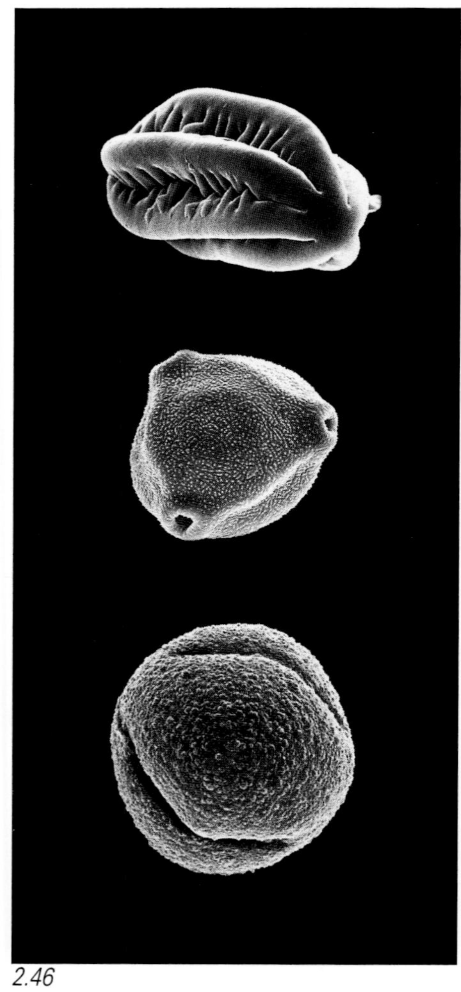
2.46

2.47

Trees and pollen

Every climate zone and time period has characteristic vegetation. The structure and density of tree-rings provide an estimate of temperature and precipitation during growth periods and an image of anomalous events. The overlapping tree-ring record now extends back about 10 000 years. In sediment cores from mires or lake, we count pollen grains from plants with warmer or colder preferences. Pollen and tree-ring studies provide quantitative indices of former climatic conditions.

2.45
Cross section through a Tamarisk pine. The form, width and irregularities of the rings register water stress during the growth periods, which can be used to estimate temperature and precipitation.

2.46
Scanning electron micrographs of three pollen grains, each characteristic for certain climates. From the top: Artemesia grass pollen for an ice age prairie, birch tree pollen for the rise of forests and oak pollen marking a climatic optimum.

2.47
Bristlecone Pine trees survive at the upper forest limit in Colorado, USA. Stumps often remain standing for centuries. Their records can be pieced together as a witness of climate over thousands of years.

Human history

The history of civilization is closely linked with the history of climate. Humans have always registered their harvest bounty. Great strides have been made in this branch of research with meticulous compilations of climate data from chronicles and diaries. The next step is to compare events on a regional, or even a global scale. A European data bank for this purpose was recently established in Switzerland.

2.48
An image of a flood event from 1562 in Zurich is overlain on a page from the weather chronicles of Paris, 1678. Two examples of how to read climate history.

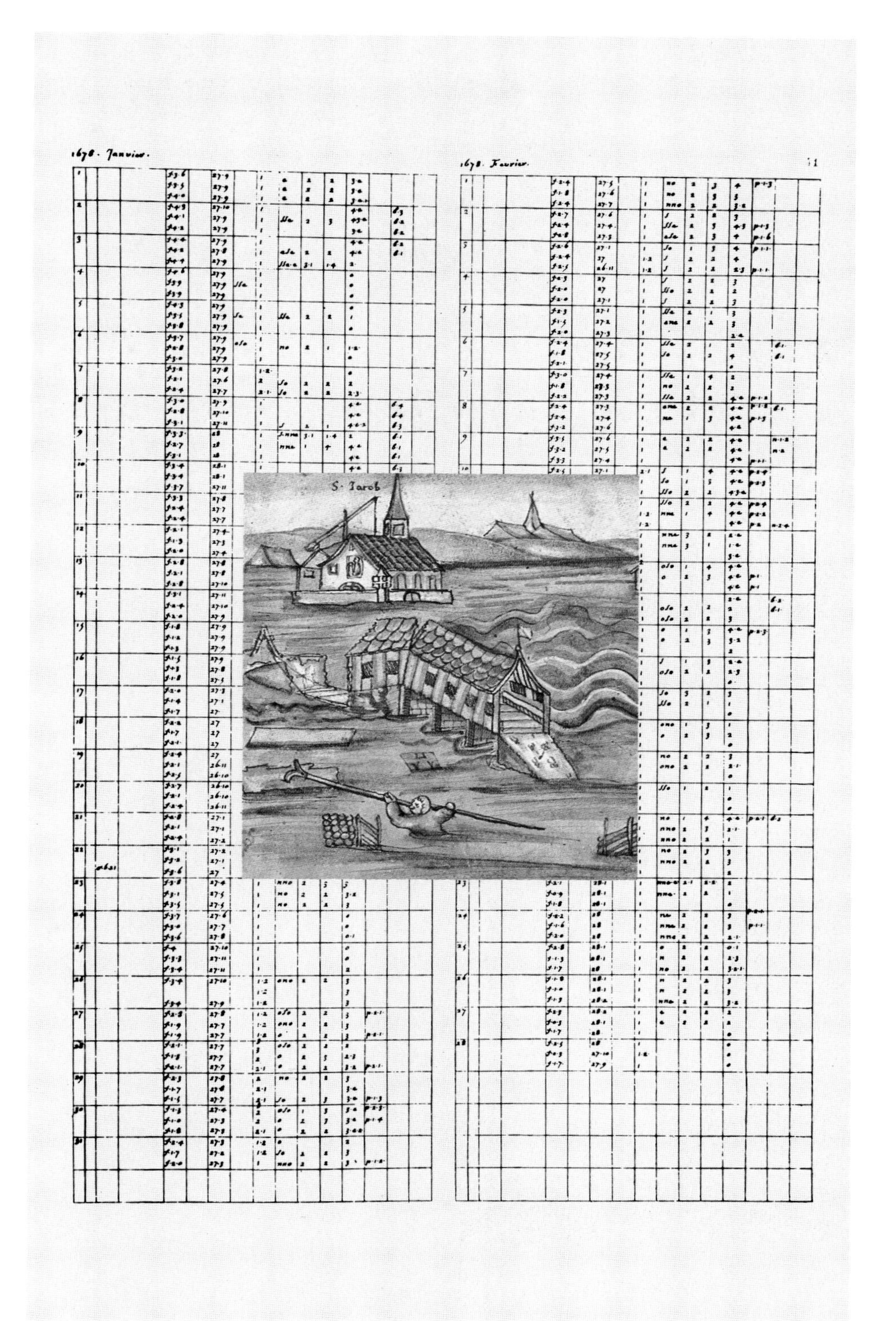

2.48

2 Ice Ages

We know now from the analyses of deep-sea cores that the great ice ages recurred on about a 100 000-year rhythm during the last million years. Massive ice sheets built up gradually over North America and Europe, but were terminated abruptly by warming trends. Each glacial stage did not remain uniformly cold; each had characteristic temperature curves. We have previously noted that the large-scale climate cycles are driven by external forcings caused by changes in the Earth's orbital parameters. There is increasing evidence that small perturbations are adequate to tip the balance for the whole, or parts of the climate system. The ocean appears to play the determinant role in this flip-flop mechanism.

2.49
Relative volumes of ice over the last 600 000 years are determined from oxygen isotopic analyses of calcareous microfossils sedimented on the ocean floor. This isotopic ratio – $^{18}O/^{16}O$ – reflects whether more or less fresh water is stored in ice caps.

2.50
The dark band in this outcrop of glacial river gravels is a lignite coal deposit near Gossau, Switzerland. It documents evidence of a warmer interval, about 40 000 years ago.

2.51
Forty-thousand years ago, a baby mammoth sank into a melting permafrost soil, died and was preserved in cold storage. This treasure, exposed by recent river erosion, allows direct studies of its former diet and, indirectly, past climatic conditions.

2.52
A famous *loess* sequence of glacial dust at Heimugon in China. A small white band, seen in the middle-left of the picture, is a 530 000-year-old soil horizon. Fossil soil profiles provide another network of pieces for our climate puzzle.

2.53 and 2.54
In 1865 Oswald Heer, a Swiss geobotanist, imagined these reconstructed ice-age landscapes, including plants and animals, for warm and cold periods of the last 100 000 years.

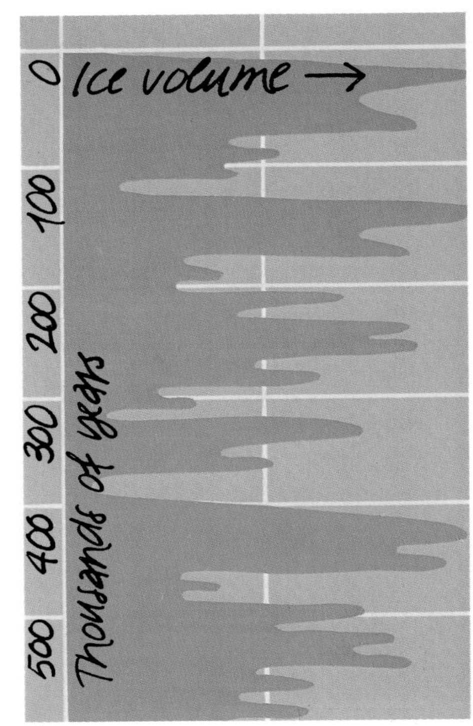

2.49

2.50

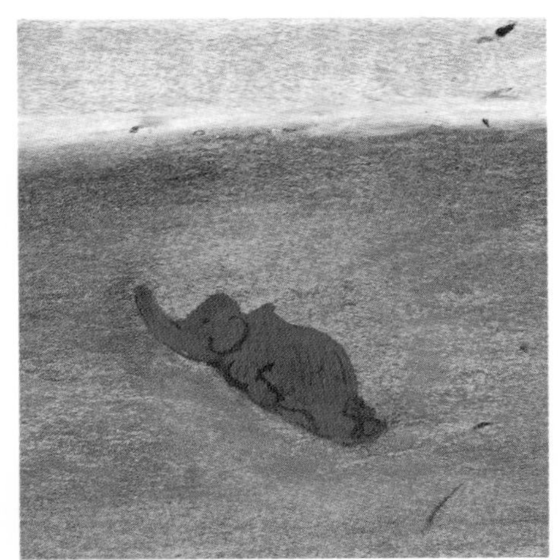

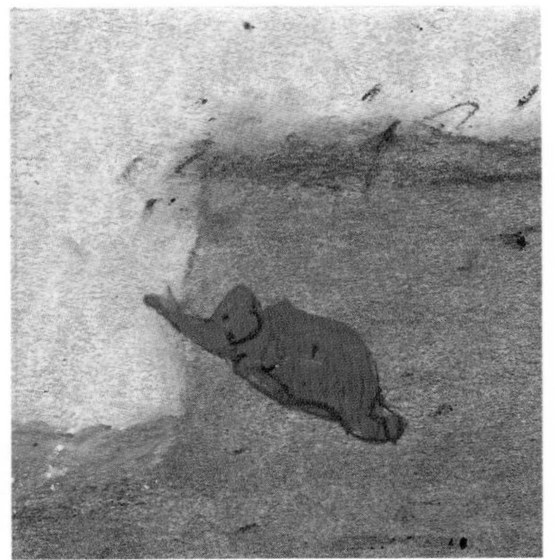

2.51

2.52

2.53

2.54

2

The Younger Dryas

About 11 000 years ago, there was a sudden setback in the general postglacial warming trend. This is called the Younger Dryas event, which lasted 800 years. Over much of Europe, the temperature sank about 6 °C within less than 100 years. The Earth's orbital parameters change too slowly to cause such effects. What happened? Scientific dectectives propose the following: Melting of the continental ice sheet of North America changes the path of the outflow from a huge glacial lake: instead of flowing down the Mississippi River to the Gulf of Mexico, it spills into the St. Lawrence Seaway and flows into the North Atlantic. Less dense freshwater builds up a layer on top of heavier seawater and changes the circulation patterns of the ocean. The Polar Front advances southward. The warm-water heater for Europe is disconnected. The return to general warming is even more abrupt than the onset of cooling. How did ecosystems of that era respond to massive, rapid changes in environmental conditions? Could they adapt? The record of the Younger Dryas event is stored in diverse climate archives such as the Greenland ice cap, or limy sediments of a small pond, like Lake Gerzen, Switzerland. Will present human impacts cause even more rapid changes, and what will be the consequences? We must support scientists to get quick answers to these questions.

2.56

2.57

2.55
Measurements of oxygen – 18 in cores from Greenland ice and sediments from Lake Gerzen, Switzerland both document the Younger Dryas cooling event. The synchroneity is confirmed by a well dated volcanic ash layer marked with the red line.
2.56
Glacier tongue into a fjord north of Thule, Greenland.
2.57
Lake Gerzen lies perched in a glacial landscape near Berne.
2.58
A model reconstruction of the presumed positions of the polar front for 5 time strips since the last ice age.

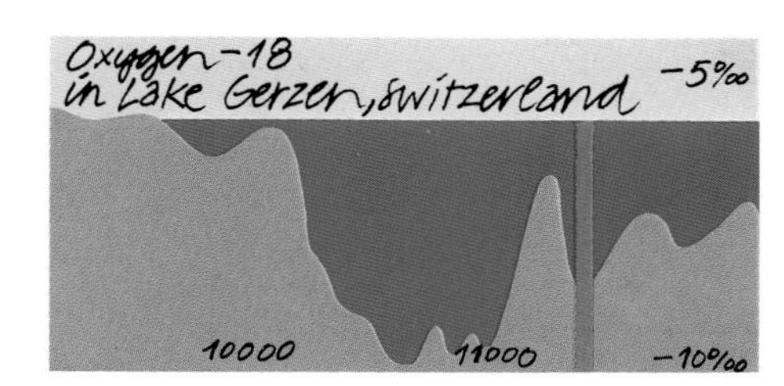

2.55

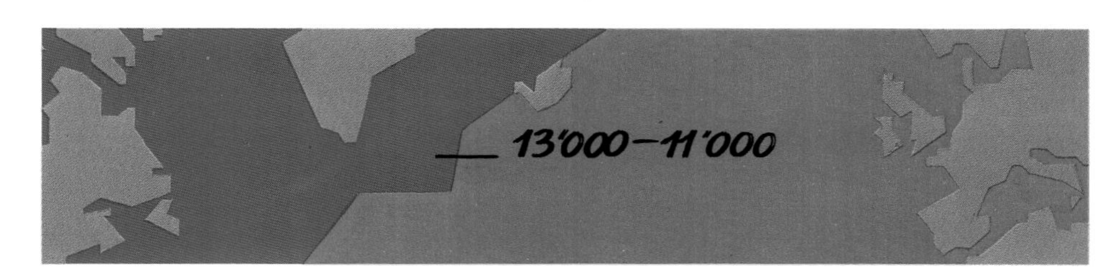

2.58

2 Little Ice Age

The term *Little Ice Age* refers to several cooling phases within the historical interval from the middle of the 13th century until about 1850. It is particularly well-documented in Europe, but has been identified elsewhere. It appears that this climate regime is directly linked to changes in solar input. The Little Ice Age was a time of reduced activity on the Sun. Sunspots were rarer. Effects on the environment are portrayed by a series of contemporary artists' impressions of fluctuations of the tongue of the Lower Grindelwald Glacier, Switzerland... a contribution of Fine Arts to Natural Sciences.

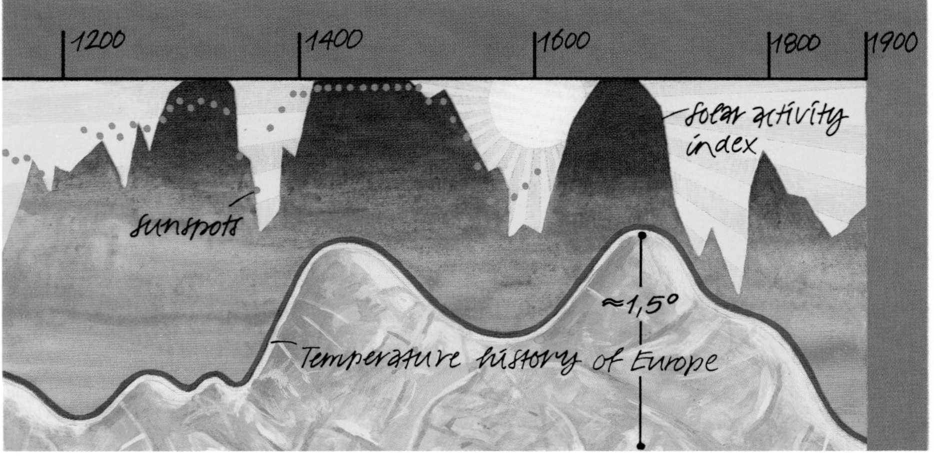

2.59
Temperature curves interpreted from proxy data for Europe suggest a direct relationship between sunspot numbers and solar activity. Red points track the intensity of the *Aurora Borealis*.

2.60
Curve for fluctuations in the position of the Lower Grindelwald Glacier tongue. Black triangles and numbers mark the picture documents.

2.61 to 2.66
The artists E. Handmann, about 1748/49; C.Wolf, about 1774/76; J. J. Biedermann, about 1808; and C.M. Lory Jr., about 1820, unkowningly portrayed climate. Photographs from 1850 and 1974.

2.60

2.61

2.64

2.62

2.63

2.65

2.66

2

The capricious nature of climate during the Little Ice Age wrought considerable social and economic hardship throughout much of Europe, particularly in Switzerland. On the other hand, tourism in Switzerland profited from the magnificent faustian landscapes created by the advanced glacier tongues. The first hotels in Grindelwald were founded after 1820; earlier travelers had to lodge in the pastor's home.

2.67
This painting from 1830 creates an illusion of the Lower and Upper Grindelwald Glaciers seen from the dining room of the Hotel Bären. The artist was improving a bit on the reality of these natural wonders; both glaciers are not visible from this perspective. The separation between the two picture windows hides the swindle of a compressed horizon.

2.67

2

El Niño

El Niño, the Christchild. Although a Christmastime event, this weather pattern brings little joy. Similar to the seven-year cycles of the biblical plagues, El Niño refers to 3–7-year recurrences of westerly winds, which drive warm equatorial waters towards the coasts of Ecuador and Peru. This inhibits the upwelling of cooler, nutrient-rich deep waters. Scientists, with the help of satellites, have found a link to another phenomenon, the *Southern Oscillation.* The SO is a teeter-totter of alternating low- and high-pressure heads centered around Indonesia, Australia and the Southeastern Pacific. Intensive ENSO events – *E*l *N*iño-*S*outhern *O*scillation – such as in 1982/83, bring droughts to some areas, while floods devastate others. ENSO also influences long-term weather development in Europe.

2.68
Time sequence of the *El Niño* event of 1982/83 in three phases: The warm Pacific waters (1) are intruded by a cool tongue of upwelling nutrient water (2). Then, the upwelling tongue is displaced by warm waters. Months later, a limited warm-water mass persists as a lid on the upwelling area (3), while parts of the Pacific become cooler than normal (4). This results in unusual weather patterns over many regions of the Globe.

2.69
Dead livestock and coastal erosion are some of the sad effects of ENSO events.

2.70
The ENSO index. Positive sea-surface temperature anomalies near Peru – in red – correlate with atmospheric pressure anomalies between Tahiti and Australia – in light blue. An example of the intimate coupling of ocean and atmosphere.

2.71
Unusually warm bathing season in 1983 along the River Aar near Berne. Does anyone here link their pleasure with causes off the coast of Peru?

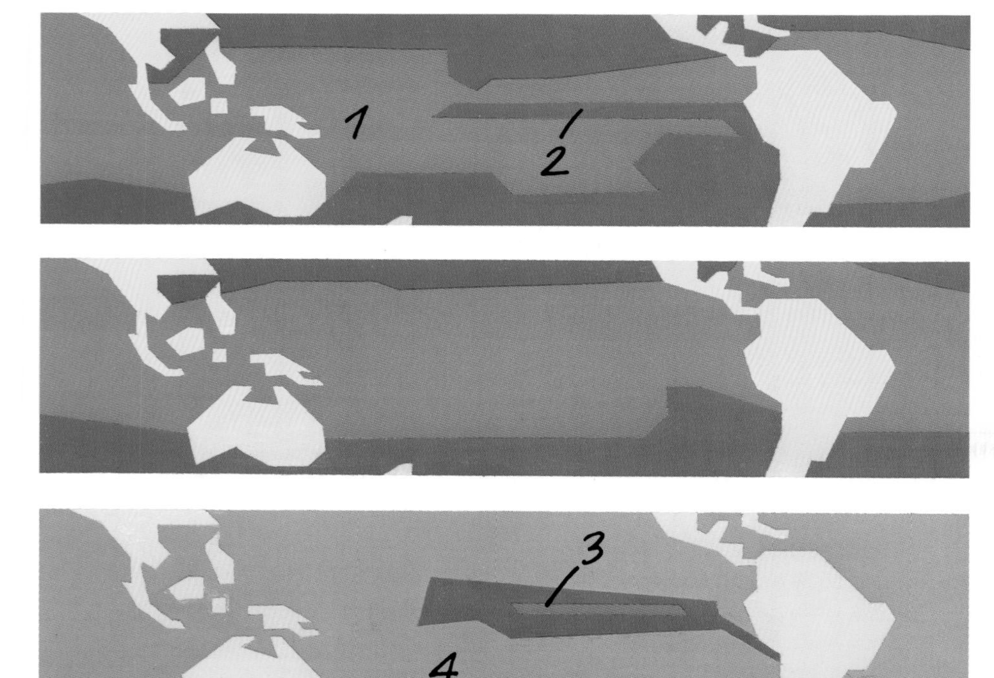

2.68

2.69

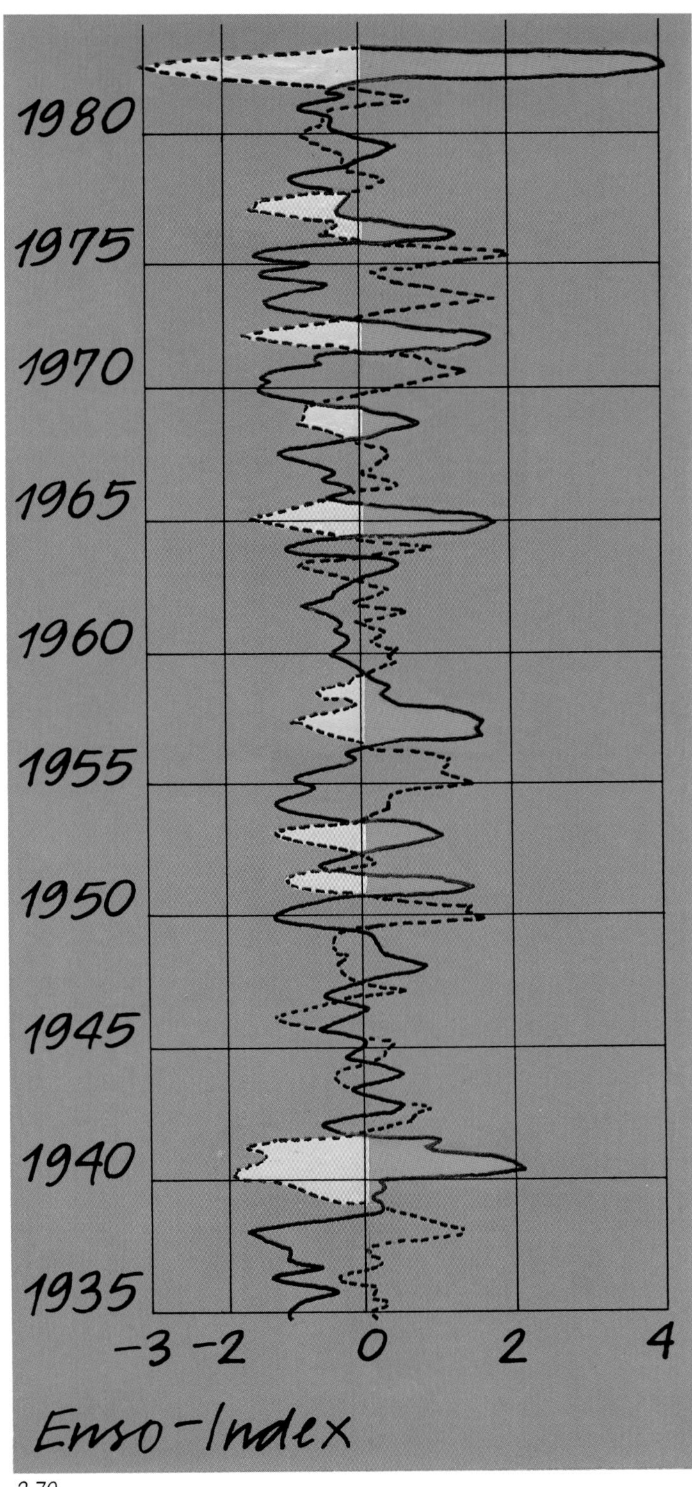
1980
1975
1970
1965
1960
1955
1950
1945
1940
1935
-3 -2
0
2
4
Enso-Index

2.70

2.71

2 What we know about climate

We are certain of the following: There is a natural greenhouse effect which keeps the Earth warmer than it otherwise would be. The Intergovernmental Panel on Climate Change, IPCC, introduces its scientific assessment with this statement.

We learn from lessons of the past. We learn from space – from our mission to Earth. We learn from hardships and benefits of daily weather. What else has this taught us about climate?

Climate varies naturally on all time scales, from hundreds of millions of years down to interannual variations. Throughout the recent history of the Earth, since humans entered this place of action and motion, 100 000-year glacial cycles have dominated climate. Global surface temperatures vary typically by 5 ° to 7 °C. Sea-level changes of up to 100 m were accompanied by large changes in continental ice volume. Natural archives like ocean and lake deposits, or ice sheets record climate extremes all over the world.

Understanding the 5 °C warming that occurred within decades at the end of the Younger Dryas cold event is important to our assessment of future human impacts on the Earth system. Scientists are on the search for high-resolution archives that can teach us about the response of paleoecosystems... but, they are still on the doorstep.

During the last 10 000 years, global surface temperatures have only fluctuated by about 1 °C on century timescales. To find examples of temperature changes similar to those predicted to occur by about the middle of the 21st century, we must look back in time more than 160 000 years. Presently, we have no paleoanalog with adequate time resolution, appropriate boundary conditions and comparable forcing of the climate system.
The major external forcing of our climate system is the effect of changes in the Earth's orbit. The response to external forcing, however, is not straightforward. Interaction between the biosphere and components of the physical climate system result in many feedback effects. Bubbles in ice help us understand how atmospheric concentrations of carbon dioxide, methane and nitrous oxide were related to temperature changes during the last Ice Age. At the moment, the time resolution of this information is too poor to study the dynamics and cause-effect relationships of changes.

Variations in the output of solar energy also affect climate. There is strong evidence that, during the medieval Little Ice Age, a quiet sun led to a mean 1 °C decrease in temperature. On a decadal timescale, solar variability and changes in greenhouse gas concentrations can lead to climatic changes of similar magnitude. Today, the anthropogenic greenhouse forcing already has the equivalent effect of a 1% increase of solar irradiance, which is much larger than existing indications of solar variability; and business as usual scenarios predict that enhanced greenhouse forcing will at least double by the middle of the next century. Furthermore, any solar change is as likely to be up as down. The hope that the Sun will balance human impacts is wishful thinking.

The ocean is the sleeping giant in the climate puzzle. It is the greatest source of uncertainty in present climate models, because relatively little is known about ocean dynamics and ocean-atmosphere interactions. The ocean is the major sink for carbon dioxide. Uptake is controlled by the physical laws of gas exchange and the rate of biospheric growth and death. In addition, the ocean acts as a buffer for some of the expected temperature increase due to greenhouse warming. Part of the excess heat is swallowed to warm the water masses. This ocean thermal sink could explain why the global temperature increase to date is only half of the expected 1 °C rise predicted by model experiments.
We do not know how changes in ocean temperature will affect circulation patterns and the role of the ocean as a CO_2 sink. Will the ocean remain a sink for atmospheric CO_2?

Climate, humans and landscape

3

Climate sculpts the surface of the Earth. Our planet has many faces: naked mountains and desert emptiness on the one hand, and an inviting, lush palette of life-supporting vegetation on the other. The ability of early humans to adapt to extreme climate conditions was their key to survival. People, animals and plants migrated into new habitats over land bridges opened by lowered sea level during the last ice age. Warming trends then submerged the bridges, stranding *Homo sapiens* on separate continents, under favorable climatic conditions. The rise of civilization accelerated human development, opening new bridges through conquest, trade and culture. Culture supplanted nature. The industrial overprint during the last century raises the question: are we the masters or guests of our planet?

3

3.1
The landscape around Lake Thun, looking southeast toward the Bernese Alps. It was overprinted by the advancing Aar Glacier; and, later, decorated by human settlement. Nearby Lake Gerzen provided a sediment history of postglacial climate change.

3.2
The landscape in the Alai Valley along the northern rim of the Russian Pamir at 3500 m helps us visualize the past landscape around Lake Thun, when glacier ice had just dissappeared and a protective film of pioneer vegetation had begun to take root.

3

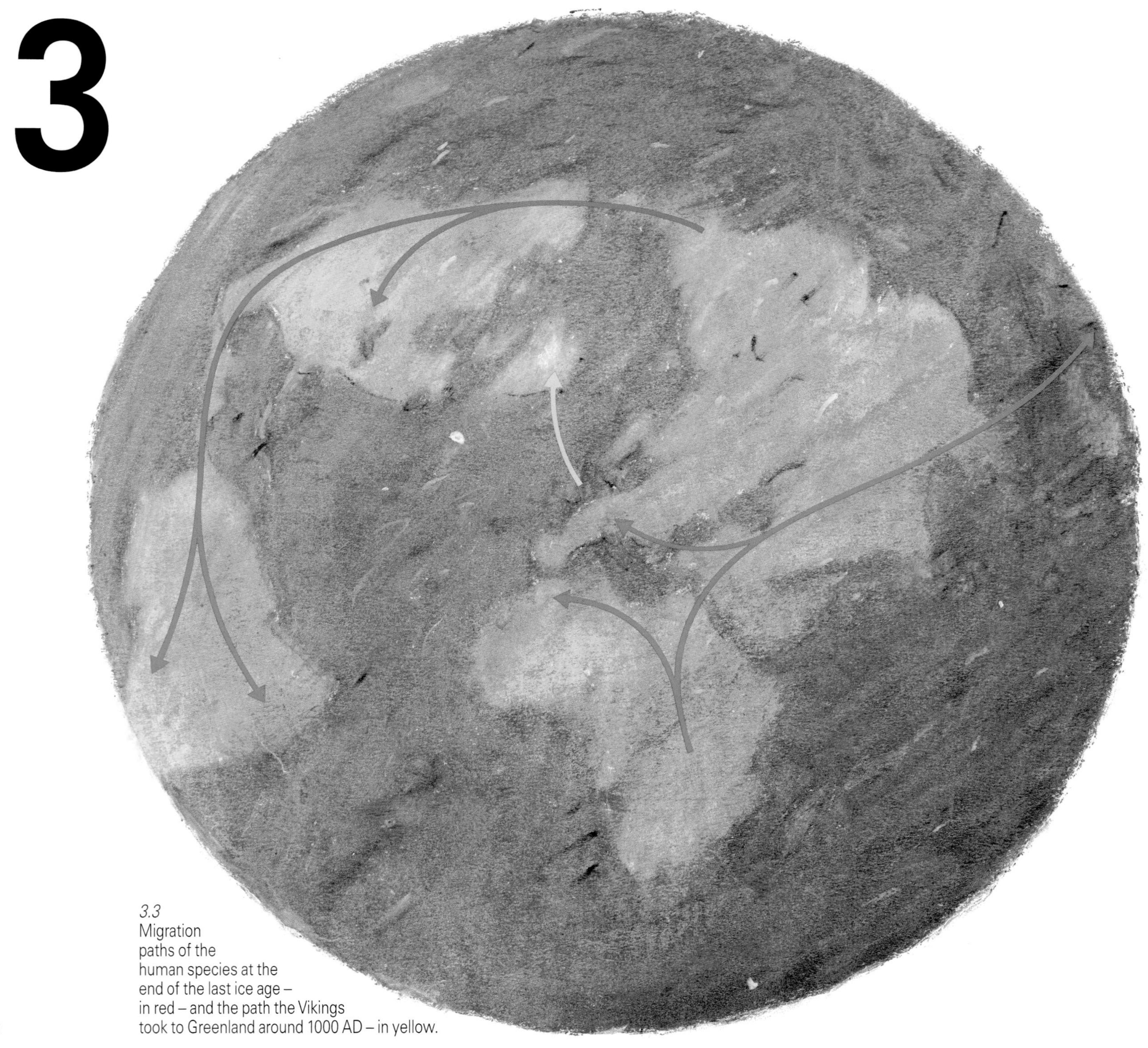

3.3
Migration
paths of the
human species at the
end of the last ice age –
in red – and the path the Vikings
took to Greenland around 1000 AD – in yellow.

Climate and habitat

Hominoids, the first human-like species, probably originated in the equatorial regions of Africa. We do not yet have a continuous evolutionary sequence linking these to *Homo sapiens*, mainly because research progress is dependent on more-or-less random discoveries. The determinant phase probably began during the last ice age when the Homo species learned to adapt to extreme climate conditions and developed a superiority over other animals. Land bridges facilitated migration. More than 25,000 years ago, hunters left Siberia, crossed the land bridge to Alaska and Canada, passed the Isthmus of Panama, and spread into South America as far as Land's End. The ancestors of the Eskimos used the same bridges to reach Greenland long before Eric the Red and his Vikings crossed the North Atlantic and landed on the world's largest island during a favorable climatic episode around the year 1000 AD. Central Africa became isolated from further contacts by the desiccation of the Sahara region. Egyptian and Syrian-Mesopotamian civilizations flourished during the *Climatic Optimum*, which was a warm interval 9,000–5,000 years ago. The early Australian aborigines also became isolated, as the sea reclaimed the Indonesian platform. The colonization of the Polynesian islands took place much later by navigation of the seas.

3.4
A land bridge between Asia and North America crosses the Bering Strait during the last ice age. The areas in light blue are now under the sea.

3.5
The Church of Hvalsey, one of the most well-preserved Viking ruins in Greenland.

3.6
Polynesians used outrigger boats to colonize the Pacific Islands. This illustration stems from a participant in a Dutch conquest of the 18th century.

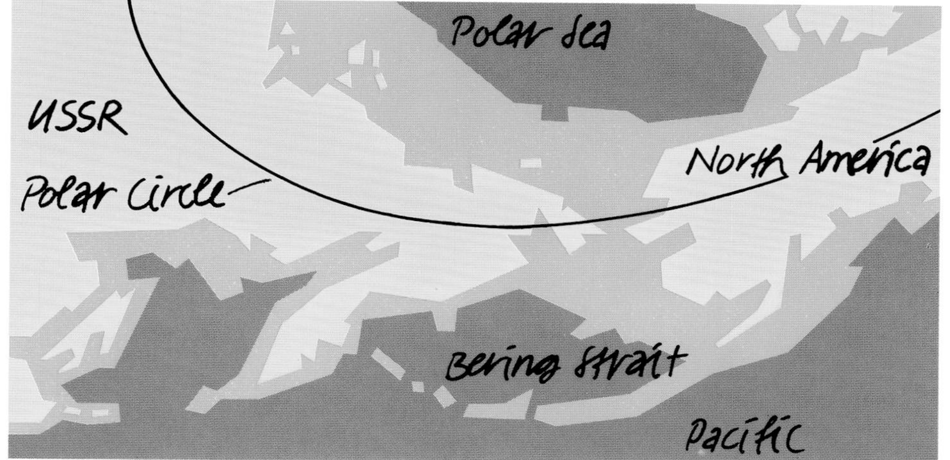

3.4

3.5

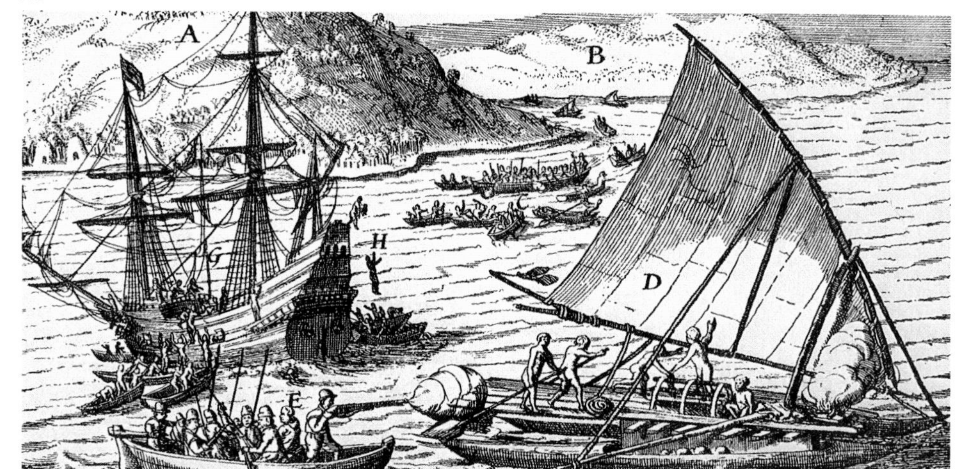

3.6

3

Climate molds landscape; landscape molds mankind – at least once-upon-a-time. The climatic stresses of the ice ages forced early communities to develop their intellectual capacity for survival. They adapted their clothing, housing and lifestyle to the natural environment. The transition from hunter-gatherer to agricultural-settlement societies amplified cultural identity and adaption to specific surroundings. This development set in motion the fascinating historical palette of differentiated cultural landscapes. We are currently taking these relationships and our landscape heritage for granted. When confronted with a deteriorating environment, can we still claim to be in tune with Nature?

3.7

Images of the Eskimo between seal skin and polyester. Today, potbelly stove heating in barracks replaces the clever cold-traps of an igloo.

3.8

And what happens in Switzerland?

3.7

3.8

3.9
Kirgisian nomads in the highlands of the Pamir. Increasingly, tents are exchanged in winter for uniform, tin-roofed shacks in collective settlements.
3.10
Petrodollars have rewritten the Tale of 1001 Nights. Air conditioners are more convenient than the meticulously designed integration of air, light and shadow that characterizes traditional mud architecture.

3.9

3.10

3 Cultural heritage since the Ice Age

Northern Africa

During the last 20 000 years, the Earth has not only been subjected to warm and cold periods, but also to distinctly arid and wet phases. The last centuries have provided a fascinating packet of data on climate history in Africa. This opens our eyes clearly to the significance of climate change for human habitability of planet Earth. During our own century, Lake Chad has varied between 12 000 and 25 000 km^2. Earlier, it was once as large as the Caspian Sea! Similarly, the Great Salt Lake was once a fresh water-body covering most of the state of Utah, USA. Vast landmasses between the equator and the Mediterranean Sea changed completely in the last 20 000 years, even without the human interference. Causes of these changes are thought to be related to shifts in the Intertropical Convergence Zone – ITCZ – a swath in the atmosphere where dry continental and humid oceanic air masses consistently merge. But what caused these shifts? Let's look into four time windows.

3.11
The expansion of the Sahara in climatic snapshots. 18 000 years ago was more arid than today. 8000 years ago was the climatic optimum, and tropical air masses supplied life-giving rain across all of the Sahara. 2000 years ago, climate trends left the region with insufficient rain. Today, population pressures and overgrazing of soils have amplified aridity effects.

3.12
Life in the Algerian Sahara.

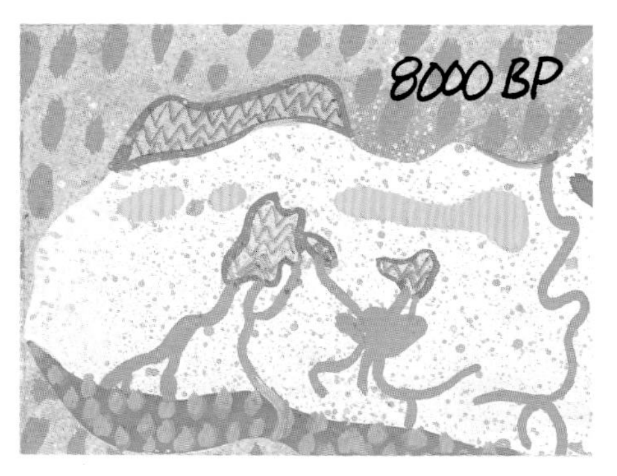

3.11

3.12

3

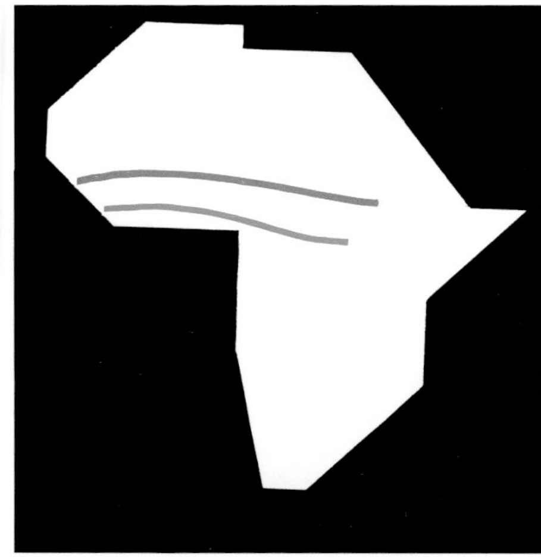

3.13
18 000 years ago, the ITCZ for summer – red – and winter – blue – is thought to have been situated further to the South. The central Saharan Mountains had a seasonally cold climate with winter precipitation. Surrounding areas were more arid than today.

3.14
Structures from ice-age snow banks in the Tibesti Mountains. The desert still conserves traces of a more-humid past.
3.15
Caravan in the Tibesti Mountains, central Sahara.

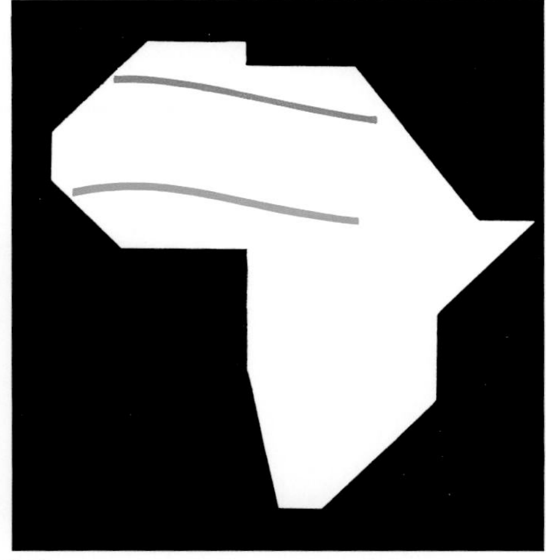

3.16
8000 years ago, the ITCZ image has changed drastically! The Sahara has abundant rivers and lakes, and continuous expanses of surface vegetation, which allowed Stone-Age cultures to settle and develop hunting, fishing and cereal economies. Impressive artifacts from early inhabitants testify to their enriched standard of living. We are still seeking better time resolution to fit these new pieces into our climate puzzle.
3.17
Sand... long ago the flourishing habitat of wild animals and humans, such as the Nubian people who carved these stones. A land of plenty provides the leisure for art.
3.18
Petrified tree stumps in the Sahara.
3.19
Changing shorelines define the fate of Lake Chad.
3.20
An oasis along the edge of the Sahara.

3.15

3.17

3.18

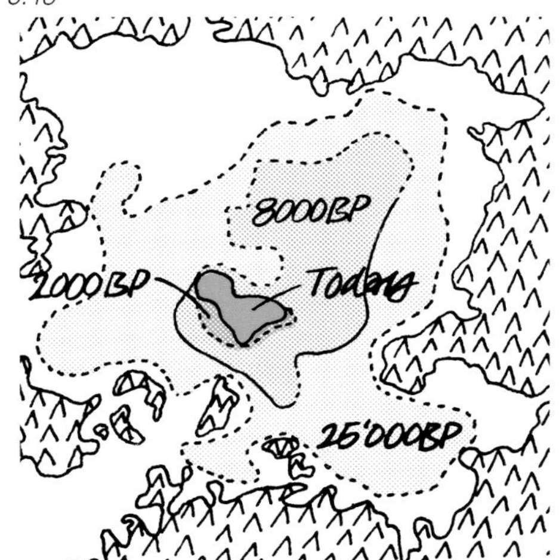

3.19

3.20

3

3.21
2000 years ago, the ITCZ is again in a new position. A stone Ramses II surveys the ruins of the downfallen Egyptian empire. Was climate stronger than Man? Surface water bodies disappeared from the Sahara about 4000 years ago, and with them, the large animals. Ecological systems collapsed. At 2000 years ago, the northern African coastal strip still remained the bread basket for the Romans, but rainfall became more irregular. Humans accelerated these natural trends by ignorant over-usage of the soils. Societies broke down. We know from lake sediments that similar processes were occurring in the regions of the Himalayas and Tibet. A causal link between Sahara and Tibet has not yet been established. The events, however, clearly high-light the fragility of the world in which we live. Are downfalls of civilizations related to climatic deterioration?
3.22
Abu Simbel. Ramses II.
3.23
Three smaller pyramids at the foot of the Great Cheops Pyramid. Originally built for eternal glorification, they return to sand.

3.22

3.23

3

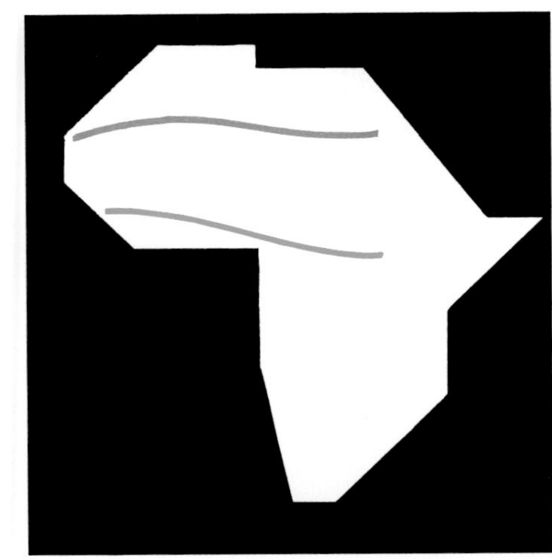

3.24
Today the position of ITCZ results in irregular rainfall. Curves document strong variability, with a distinctly downward trend in average annual rainfall in northern Africa. The last two centuries witnessed severe drought periods in the African ITCZ. Exploitation of fossil groundwater to counteract the droughts has increased. This groundwater is left over from the earlier humid phases. How long will these reserves last? Can we expect them to refill?
3.25
The hydrologic cycle in an arid zone. Arrow widths are proportional to transfer amounts. Only a small portion of rainfall recharges the groundwater. Increased withdrawal could upset the balance.
3.26
Graph showing the decreasing trend in annual rainfall received in Mali.
3.27 and 3.28
Deep wells drilled into aquifers containing fossil water quench a thirsty populace, as here in Niger.
3.29
Traditional irrigation methods in the Tibesti Mountains attempted to stretch the precious raindrops available.
3.30
. . . whereas modern techniques pump groundwater in the Libyan desert. Food for a growing population is produced with the help of giant rotary irrigation spindles.

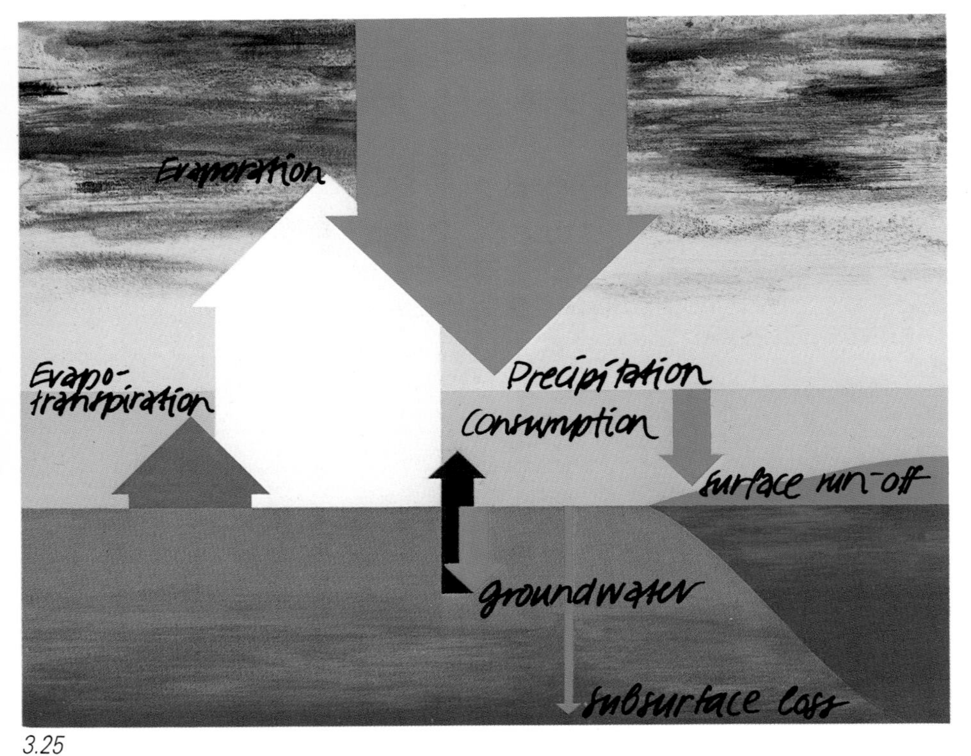

3.25

mm
800 Precipitation in Mali (500–900 mm regime)
600
400
1940 1950 1960 1970 1980

3.26

3.27

3.28

3.29

3.30

3 Climate and Europe

We know that tree rings and glacier fluctuations chronicle climate history. Trees react each year to environmental variations. Glaciers react with time lags of up to decades, depending on their size. Careful study of both patterns reveals surprising correlations and links to our own history. *Hannibal's* crossing of the Alps seems to have taken place in the nick of time, before a climatic deterioration which was registered as an advance of the Aletsch Glacier. It is said, that he also brought wine grapes to the alpine region. Conspicuous improvements in climates of the 12th, 14th and 17th centuries allowed *vineyards* to flourish at higher altitudes. The *Walser* people, a hearty mountain folk, had to evacuate dry, southern alpine valleys and cross to the North. There, valley pastures were already occupied, and they were forced to cultivate slopes near the forest limit. Their isolated homestead patterns characterize the alpine landscape along the Swiss-Austrian frontier.

3.31
The Aletsch Glacier today.

3.32
Above, 1500 years of temperature and precipitation information obtained by the study of tree-ring series from northern Europe. Below, moraine evidence for fluctuations of the Aletsch Glacier, Switzerland, extend the record back a thousand years more. Green stripes represent dated trees, which tell when areas were ice-free, and, later, over-ridden by glacier advances.

3.33

3.31

Tree-ring curve Lappland
Warmer △
Colder ▽
-500 -400 -300 -200 -100 0 100 200 300 400

Retreat △
and Advance ▽
of Aletsch Glacier, Switzerland
?

3.32

3.34

3.35

3.33
Hannibal's elephants crossing the Alps during a favorable climate episode, 218 BC.

3.34
This section of a medallion illustrates the favorable wine-growing conditions in Europe during the 17th century. Vineyards endangered cereal crops to such an extent that strict laws were issued for their regulation.

3.35
A homestead near Davos, typical of the Walser people .

600 700 800 900 1000 1100 1200 1300 1400 1500 1600 1700 1800 1900 Today

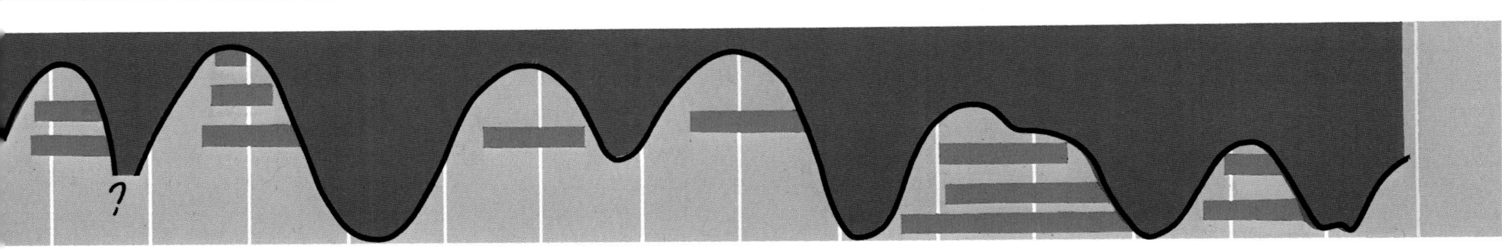

3

Climate probably sealed the fate of the *Silk Road*. This famous route between the Roman and Chinese empires experienced peak activity during a favorable period before the end of the first millenium. Trade created bustling cities along the perimeter of the desolate Tarim Basin in the center of Asia. These cities were slowly buried by the shifting sands of the Taklamakan Desert which also hindered Marco Polo in his travels between 1271 and 1295. Today, we believe that the desiccation of the Tarim Basin is directly related to retreating glaciers in the surrounding Tienshan, Kunlun and Pamir mountains. The basin receives almost no rainfall. Diminishing glacier meltwater cut off the springs feeding *Qanats*-underground irrigation canals.

3.36
Venice – point of departure for Marco Polo.
3.37
Mountain panorama of the Russian Pamir. One of the routes of the Silk Road led through the Alai Valley.
3.38
The Temple of Heaven in Beijing.
3.39
The Genoese World Map of 1457 was an important basis for trade routes in the Middle Ages.
3.40
This close-up of the Great Catalan World Atlas of 1375 shows a trade caravan passing through the Taklamakan Desert, central Asia.

3.36

3.37

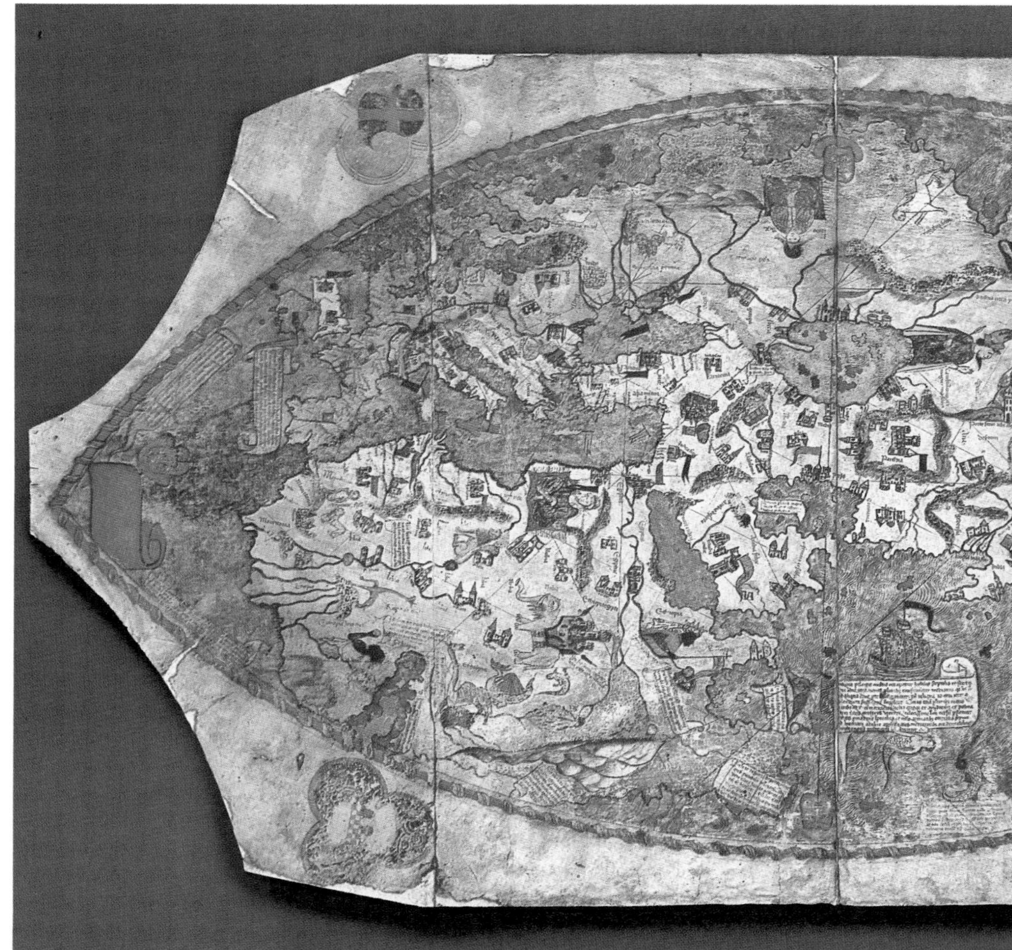

3.39

3.38

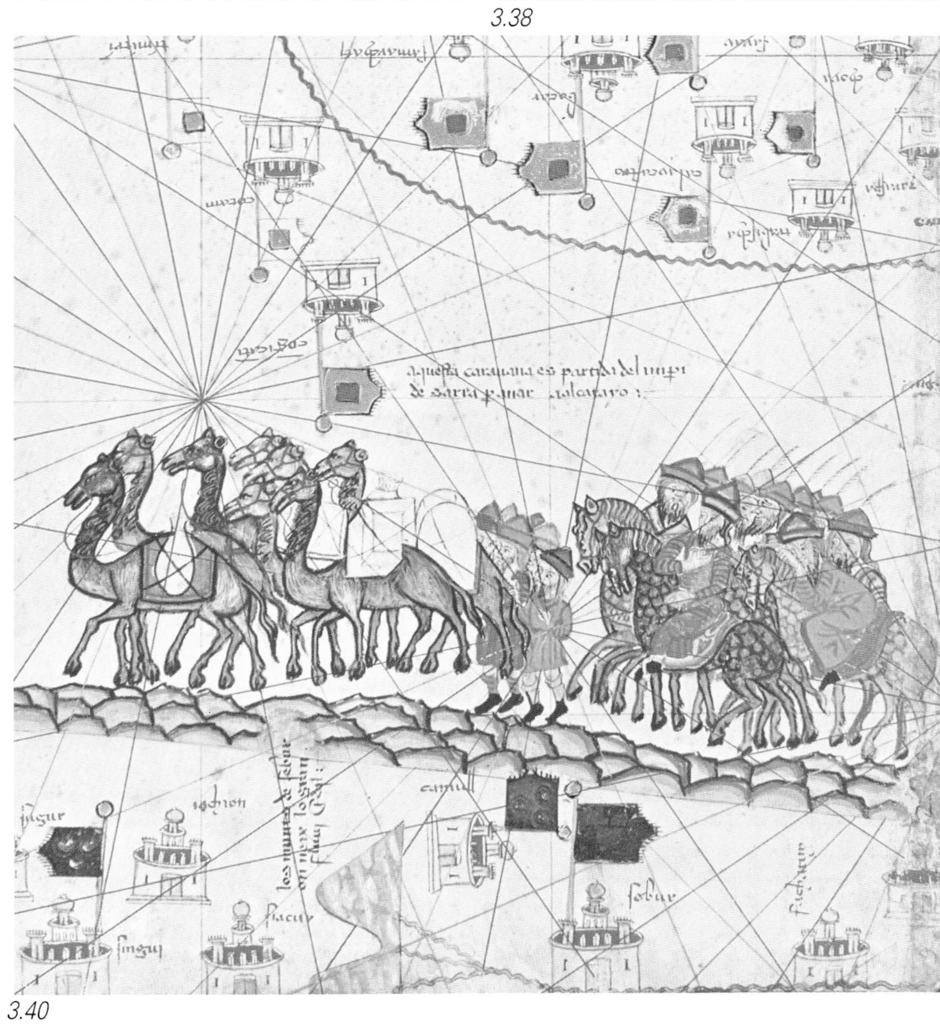

3.40

3

Climate and society

The last 800 years witnessed a Middle Ages warm period, followed by the Little Ice Age previously described. This pattern does not imply a single, long warm period followed by a single cold period. Rather these general trends are punctuated by abrupt changes between warm, wet, cold, or dry conditions. Although the long-term average temperature did not exceed 1 °C, the climate periods had a strong influence on the social structure in Europe. The people lacked our moderating technical amenities. They were subjected directly to the unforeseen moods of weather and its consequences. The Great Plague of 1348–1350 claimed more than one quarter of the population in Europe. This epidèmic was preceeded in 1342–1347 by the wettest and coldest summers of the last millenium. Following the Great Death, the Swiss Cantons of Zürich, Glarus, Zug, and Berne joined the Confederation Helvetica between 1351 and 1353. Was this climate or coincidence?

People associated extreme climatic events with supernatural causes. These beliefs still survive in traditions. Climate history profits from calamity. In the bitter cold winter of 1573, a procession from Münsterlingen marched across the frozen Lake Constance. They carried a colored statue of the holy disciple St. John to display in the city of Hegnau. Tradition required its return via the frozen lake. This did not occur until 100 years later. The last procession for St. John was in 1963.

Science matured and focused more and more on describing Nature's experiments. Naturalists such as J.J Scheuchzer, H.B. de Saussure, F.J. Hugi, or L. Agassiz sought answers to the puzzles of the alpine world and the ice ages. Advancing and retreating alpine glaciers offered a playful outdoor classroom. Elsewhere, society innocently began to influence Nature's experiments with, at first, only local effects.

3.42

On February 2, 1830, Burghers from Hegnau once again chill their feet behind the bust of St. John, crossing frozen Lake Constance. The lake froze 33 times between 875 and 1963, but over half were during the Little Ice Age, between 1378 and 1573.

3.41

"The Plague" by the Swiss painter Arnold Böcklin, who lived 1827–1901.

3.42

3.43
F.J. Hugi and friends,1830. Swiss scientists attend a natural classroom near the Jungfrau Massif.

3.44
The 19th century environmental problems, such as those of Paris shown here, were viewed only in terms of their social effects. Such problems now approach epic proportions in major world cities. Did we play hooky from one of the lessons of the Past?

3

An epoch of swooning alpine romanticism dawned as a possible reaction to city squalor and social tensions. Did the artists attempt to improve on reality?
In 1779, Goethe visited the Staubbach Waterfall near the Jungfrau Massif. This scene inspired his "Song of the Water Spirits"

"The human soul
Resembles water
From heaven it comes
To heaven it goes
And again to the Earth
Eternally cycled..."

In a letter, Goethe relates that the Staubbach Falls symbolized the highest levels of tranquility; the falling mist had overwhelmed him with a veil of serenity. Fears of mountain treachery subsided with the Enlightenment of the 18th century.
Society began awakening in the 19th century to their own environmental impacts.

3.45
The Staubbach Falls around 1832.

3.45

The awakening

Until the beginning of the 19th century, hungry peasants could not flee the destitution caused by large-scale harvest failures. New frontiers of western America and the rapid and cheap transportation by steamships opened new perspectives. No longer submit and adapt; rather escape! The slogan called for new shores. Agricultural crises lasting from 1845 to 1855, and again around 1880 drove millions of Europeans overseas. Enthusiastic letters of quick successes from the pioneers, with money sent for relatives, soon convinced hesitant drifters to turn their backs on the old country. The Industrial Revolution imposed a new social order and, with it, new requirements for a rapidly growing population. Perhaps communal dwellers developed a new awareness for air quality once the first belching factories brought aggravating odors into their neighborhoods. Fresh air in the Alps was once held responsible for idiocy and stubbornness in its folk; there is even a medical dissertation from the early 18th century to prove it. It maintained that the air in the Swiss Alps was as dangerous as the air in the Tyrolian Alps. Professional opinions have changed. The positive effects of alpine air have become a trademark.

3.46
Images from a contemporary humor magazine show how Mother Helvetia pulls the livelihood from under her children. Disappointment and suffering await abroad.

3.47
Graph showing the quantity of people that emigrated from the Old World below the quantity of foodstuffs imported from the New World. Green bars mark the years of poor harvests in Switzerland. The red curve is the relative number of Swiss emigrants; blue, for the relatively smaller proportion for Europe as a whole.

3.46

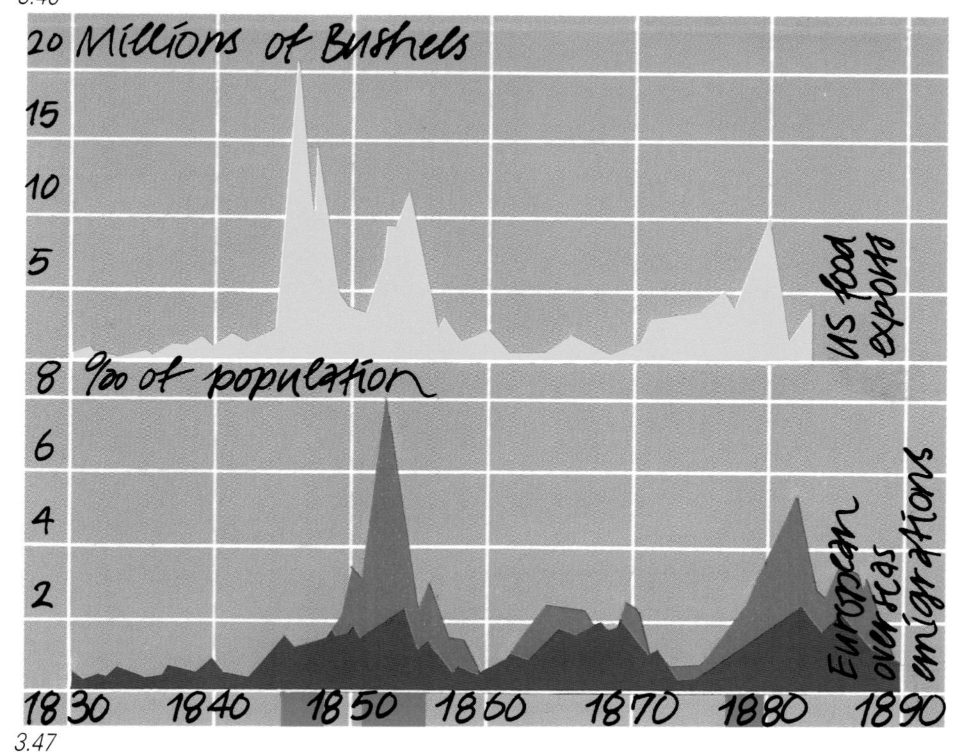

3.47

3

Something akin to an environmental consciousness arose in response to real problems. A Swiss forest supervisor reported in 1849 that "Nowhere has the clearing of forests in the Canton of Berne progressed so far as in the Emmen Valley; and thus it is plagued by great damage from floods and damming of the streams." Legislation to stop deforestation and encourage replanting passed only after catastrophy threatened survival. Can we use 19th century solutions to protect a global forest today?

3.48
Ferdinand Hodler once produced a version of his famous logger's pose for a Swiss 50 franc note. The symbolism is less appropriate today, when we are paying to keep the forests alive.

3.49
Catastrophic flooding of 1853 in the Emmen Valley; a schoolhouse in Wasen collapses.

3.48

3.49

3.50
Much smoke and belief in unbridled progress covers sparks of environmental conscience.
3.51
The CO_2 odyssey begins.

3.50

3.51

3 Climate, humans and landscape

Climate sculpts the face of the Earth. But more and more, culture and industrial activities leave their mark on the landscape. Predicting how humans, the water cycle and ecosystems will respond to global warming and what climate feedbacks they will produce is a difficult task.

As long as precipitation is adequate and the rate of change is slow, a warmer climate has some potential benefits. However, lessons from the past teach us how suddenly the Earth system can respond in step-like changes. Climate change, therefore, will also have many negative impacts on society and especially ecosystems. Ecosystems will respond to local changes – and rates of change – in temperature, precipitation and soil moisture, as well as to the frequency of their extremes. In principal, rising atmospheric CO_2 levels can increase the rate of photosynthesis of some plants, because the CO_2 fertilization results in more efficiencient use of water, light and nutrients. However, not all plants respond similarly to high CO_2 levels and changes in other climate variables. Hence, ecosystems will change in structure and composition, and this will change our familiar local environments.

Current models are unable to make reliable estimates of meteorological parameters on the required local scales. Regional scenarios, however, are possible.

The IPCC reports results of high-resolution model simulations of a global mean warming of 1.8 °C by the year 2030. In *Central North America*, the warming varies from 2–4 °C in winter and 2–3 °C in summer. Precipitation increases range from 0–15% in winter, whereas there are decreases of 5–10% in summer. Soil moisture decreases in summer by 15–20%. In *Southern Asia*, the warming varies form 1–2 °C throughout the year. Precipitation changes little in winter, and generally increases throughout the region by 5–15% in summer. Summer soil moisture increases by 5–10%. In the *Sahel*, the warming ranges from 1–3 °C. Mean precipitation increases and mean soil moisture decreases marginally in summer.

In *Southern Europe,* the warming is about 2 °C in winter and varies from 2–3 °C in summer. There is some indication of increased precipitation in winter, but summer precipitation decreases by 5–15% and summer soil moisture, by 15–25%. In *Australia,* the warming ranges from 1–2 °C in summer, and is about 2 °C in winter. Summer precipitation increases by around 10%, but the models do not produce consistent estimates of the changes in soil moisture. The area averages hide large variations at the subcontinental level. Although the confidence in all of these estimates is low, they suggest how the Earth could change.

Unforeseeable regional and local changes are casting a cloud of uncertainty, not only over our future landscape, but also over future food production, which must supply a rapidly increasing population. Growing populations tend to overcrowd cities. In the 18th century, 3% of the world's population lived in urban areas; today, 40% are in cities and in 2025, cities will house 60% of the entire population. Nearly all of this future growth will take place in developing countries, resulting in severe societal and environmental problems and affecting the quality of life. Since the majority of densely populated areas are exposed to natural hazards like flooding, drought, landslides, or storms, extreme weather events will put large populations at risk. This could initiate massive human migration, leading to social instability.

Until now, our human landscape has been a Sleeping Beauty. Will the kiss of mankind bring life or death?

People – Climate

4

"Be fertile, go forth, multiply, and make the Earth thy servant." This phrase from Genesis seems increasingly to be one of the big misunderstandings between Man and Nature. Long have we worked to modify the natural environment according to our own definition of progress. Yet our multiplication, out of control, clearly exposes the delicate relationship: Humans-Nature. We are interfering in almost all of the natural cycles without sufficient knowledge of the rules of the game. Our meddling with the climate system, mainly by the burning of fossil fuels and the production of CFC's, methane and nitrous oxides, is becoming a global challenge for human societies.

4 Population grows- thirst for energy grows

Global population has tripled in the last one hundred years; doubled within the last thirty-three. In 1986, it surpassed 5 billion; in the year 2000, it will be 6 billion. How should it – how can it – go on? Can we stabilize, the world population at 8–11 billion as optimistically predicted by United Nation agencies? Population growth is strongly non-uniform with respect to individual countries and continents. Population growth is greatest in Africa, Latin America and Southeast Asia. The mismatch between people and resources, already precarious for foodstuffs, widens. When appetites are satisfied, people clamour for a higher standard of living and thirst for more energy grows. The magic formula for energy is to burn; first wood and biomass, then more and more of the fossil fuels: coal, oil and natural gas. We produce about 85% of our primary energy by burning processes, which release carbon dioxide, CO_2, and water as end products. We long considered the fossil fuels as unlimited gifts of Nature, at the disposal of our civilization. Ready access made us complacent. Suddenly, the blessing seems a curse. Without realizing it, we have irreversibly begun an experiment with the sensitive regulatory mechanisms of our global climate. Will our experiment become a nemesis?

4.1
Growth of world population.

4.2
Growth of fossil CO_2 emission rates on a logarithmic scale. Maximum increases followed World War II. Can we change the trend within the next generation?

4.3
As population grows, cultivatable land declines.

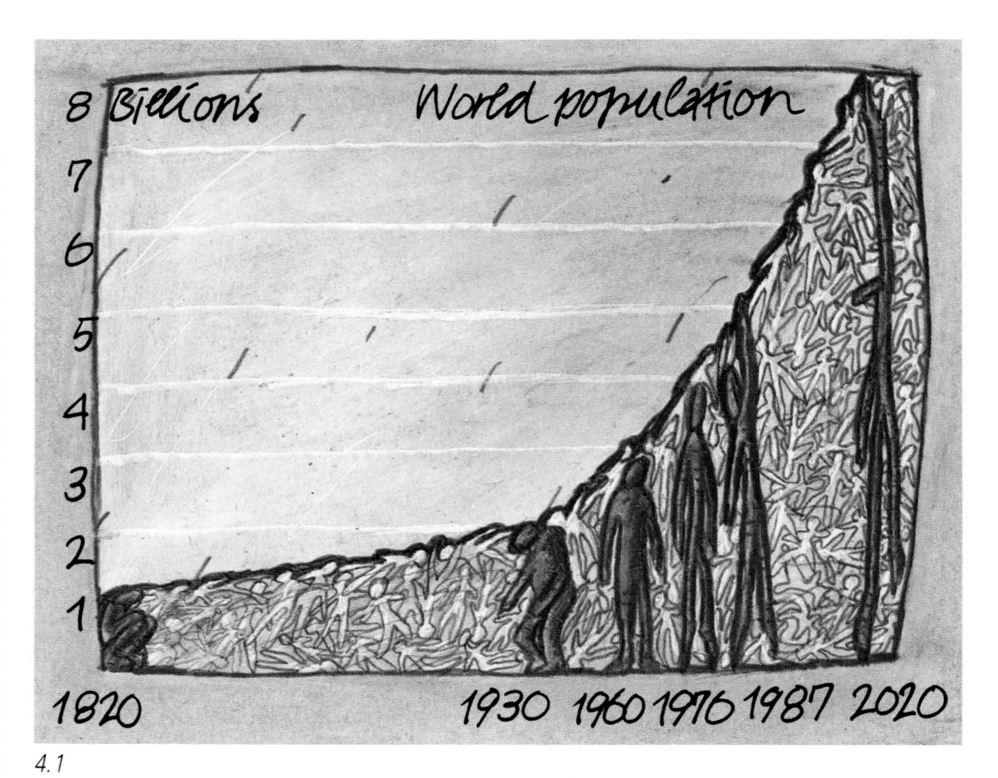

4.1

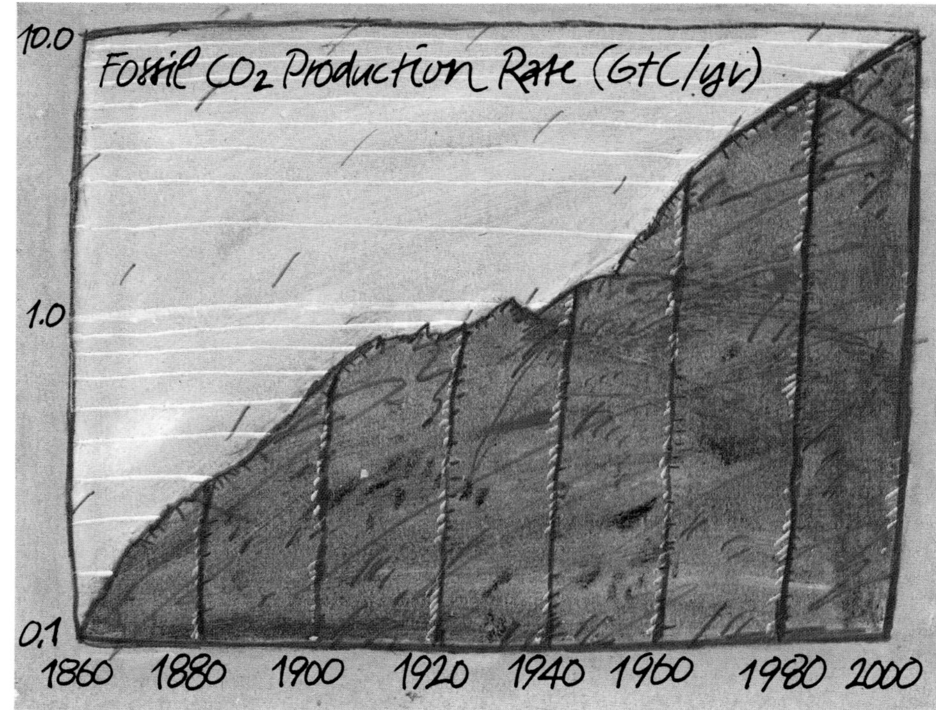

4.2

4.3

4 Cities become crowded

The comfort of the rich is a goal for the poor. People crowd into cities, believing that dreams can be fulfilled. Cities in developing countries display sprawling growth. These rates outpace any form of city planning and undermine all rational use of energy. Traffic, waste and water crises become routine. Smog alarms are only possible where ordinances and instruments exist. One fatal characteristic of humans is that we seem to learn only from our own, often bitter, experiences, rather than follow history's lessons.
And in good, old Europe? Industralizied countries are beginning to see the limits. By necessity, people realize that fewer cannot simply occupy more and more living space. Everyman-his-own-castle will plaster the landscape in concrete.
Is that enough?

4.4 to 4.6
Nairobi around the turn of the century, during the thirties and today. A chronicle from drained colonial swamps to a bustling conference center. This symbol of rising national identity in Africa is confronted today by the massive influx of rural populace.

4.7
Idyllic scene along the Lake of Neuchâtel around 1830.

4.8 and 4.9
The same view, just after WWII, and today? Concrete has taken a prominent place, not always as landscape improvement. Growing public awareness has at least spared the region from skyscrapers.

4.4

4,5

4.6

4.7

4.8

4.9

4 Greenhouse gases increase

The concentration of natural greenhouse gases has remained uniform over the last thousand years. The concentrations began to rise with industrialization. The rates of increase have risen sharply in the last decades and overprint natural variations.

4.10
Evolution of four greenhouse gases in the atmosphere over the last 250 years, and their relative contribution to man-made greenhouse effects for the decade 1980–1990

4.11
The famous curve of rising atmospheric CO_2 measured on Mauna Loa, Hawaii. An end is not in sight.

1957

Human beings are now carrying out a large scale geophysical experiment of a kind that could not have happened in the past nor be reproduced in the future. Within a few centuries we are returning to the atmosphere and oceans the concentrated organic carbon stored in sedimentary rocks over hundreds of millions of years. This experiment, if adequately documented, may yield a far-reaching insight into the processes determining weather and climate.

Roger Revelle and Hans E. Suess.

1990

We are certain of the following:
There is a natural greenhouse effect, which already keeps the Earth warmer than it would otherwise be. Emissions resulting from human activities are substantially increasing the atmospheric concentrations of the greenhouse gases: carbon dioxide, methane, CFC's, and nitrous oxide. These increases will enhance the greenhouse effect, resulting on average in an additional warming of the Earth's surface. The main greenhouse gas, water vapor, will increase in response to global warming and further enhance it.

Intergovernmental Panel on Climate Change

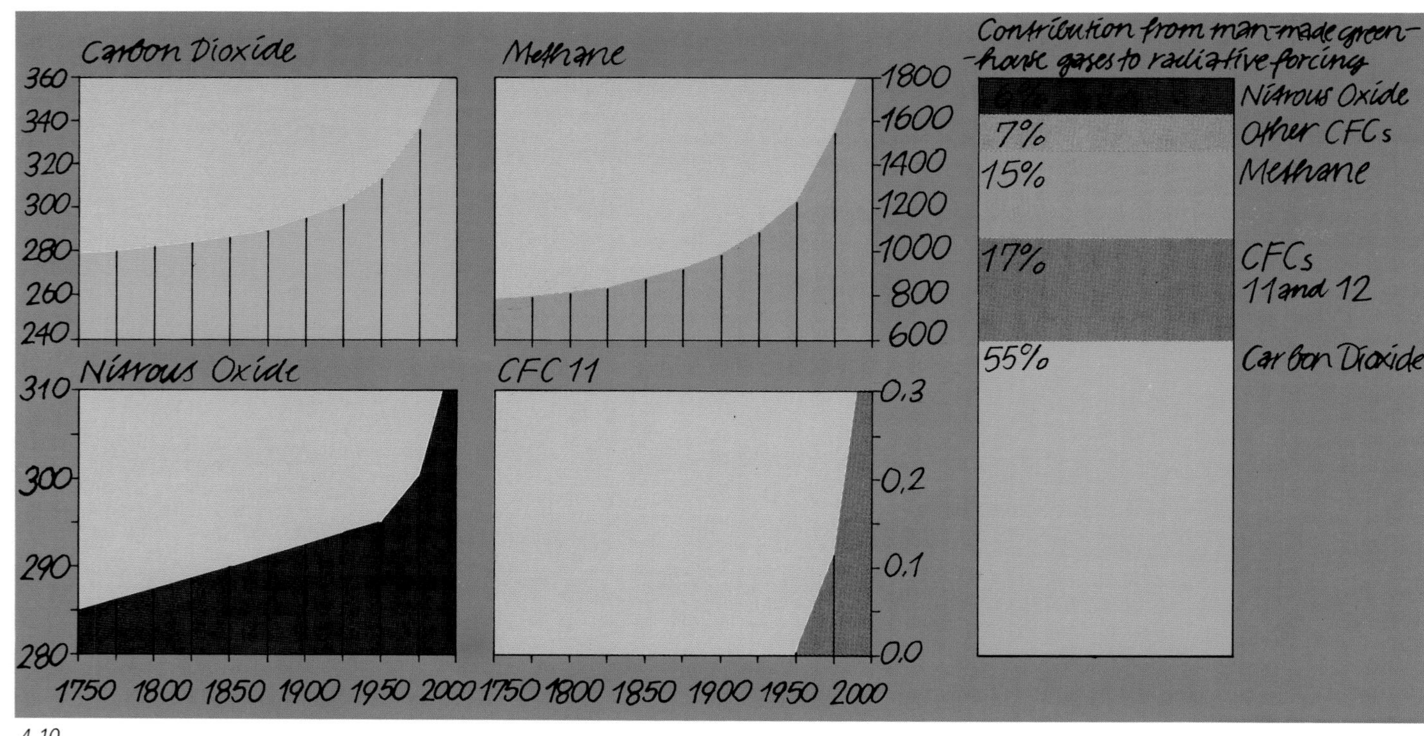

4.10

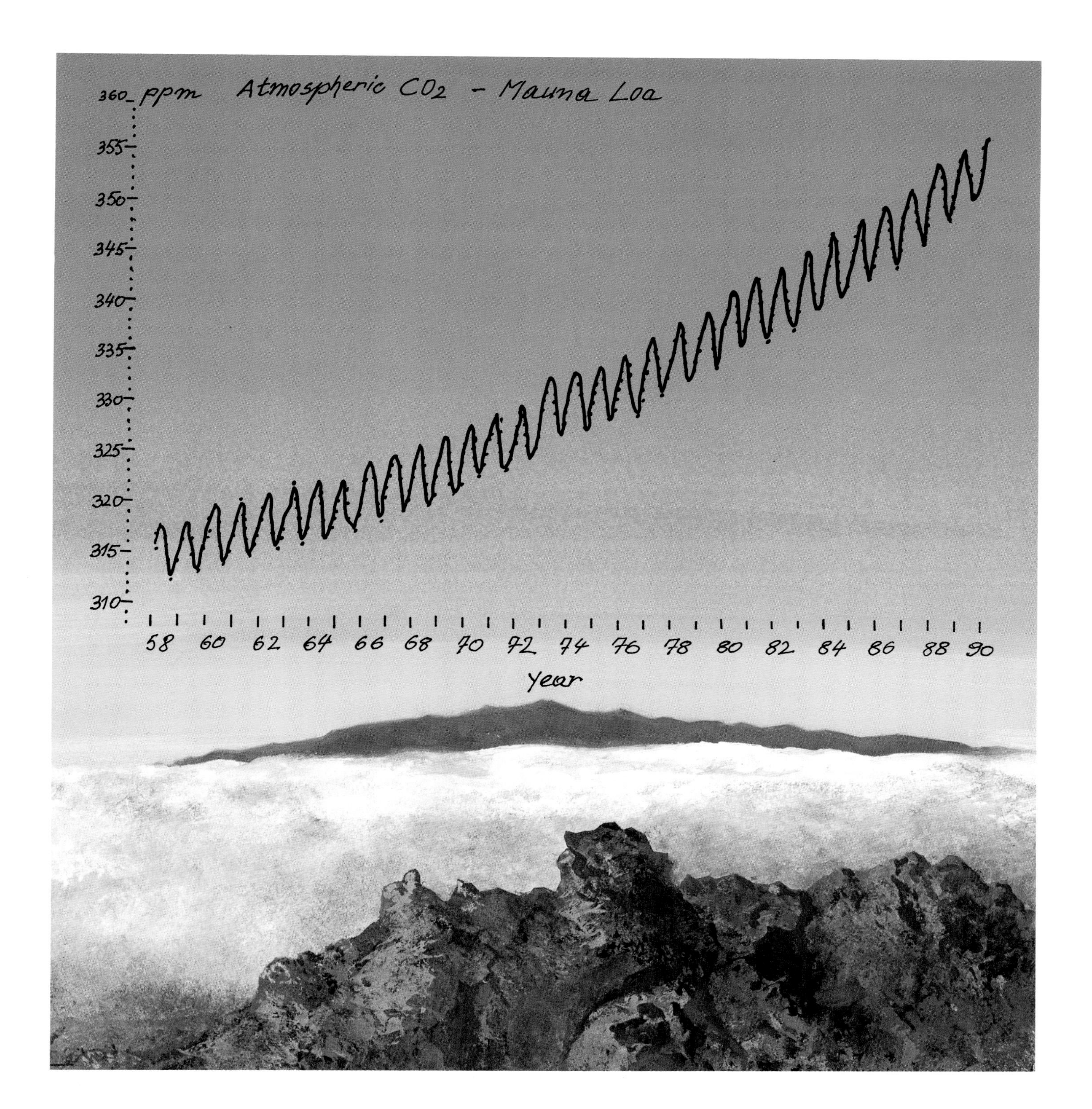
ppm Atmospheric CO2 - Mauna Loa
360
355
350
345
340
335
330
325
320
315
310
58 60 62 64 66 68 70 72 74 76 78 80 82 84 86 88 90
Year

4

Carbon dioxide

For years, the causal link between global climate and the atmospheric carbon dioxide concentration was not accepted. This ostrich behavior conforms to human nature, but is a dead end. There was no doubt about the measured increase of CO_2 in the atmosphere since the 1950's, but what was it like earlier? Was the CO_2 increase measured at Mauna Loa due to fossil fuel burning, or was it just one of Nature's moods? The answer is frozen in ice. Air bubbles locked into cold polar ice caps contain an unadulterated document of past atmospheres. Research teams from France and Switzerland each conquered experimental difficulties to reconstruct a record of atmospheric carbon dioxide concentrations in the past. These results prove that today's CO_2 increase is irrefutably linked to Man's activities. These data also confirm that CO_2 and climatic fluctuations, alarmingly, changed hand in hand during the last glacial cycle.

4.12
The climate archive of Antarctic ice.

4.13
Left: CO_2 values from ice cores – in light blue – join smoothly with the recent curve of atmospheric measurements – blue. Right: Ice cores up to 160 000 years old prove a link between CO_2 and climate.

4.14
Air bubbles in ice contain samples of ancient atmosphere.

4.12

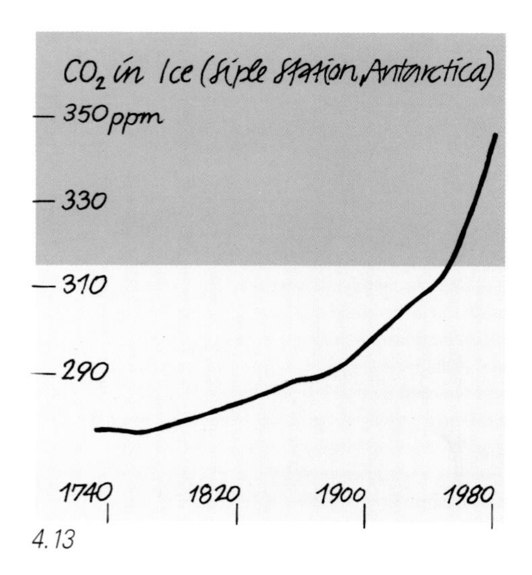

4.13

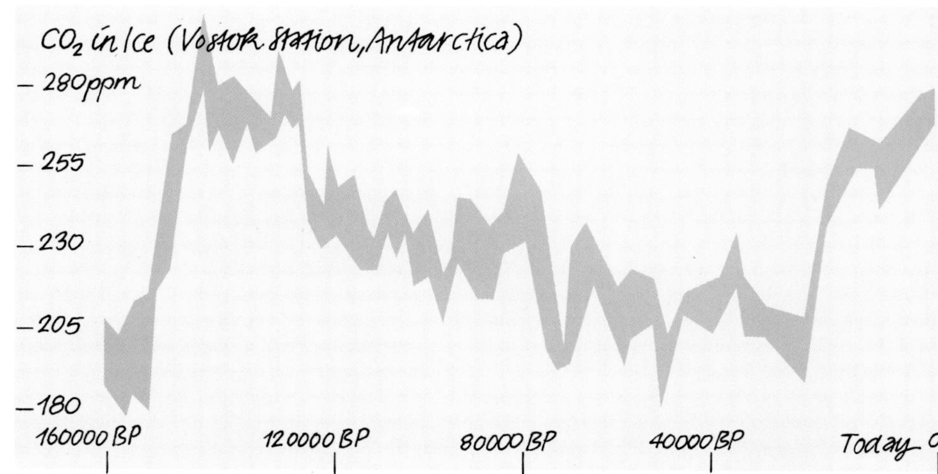

4.14

4

Methane

Everyone knows swamp gas. Bubbles of methane rise in stagnant ponds as organic matter rots. Methane, CH4, is one of the trace gases contributing to the greenhouse effect. Its concentration today is double that of 200 years ago. Not by coincidence, methane grows with the world population. One of the causes is increasing production of rice and meat. Bacteria in rice paddys and in the digestive tracts of the cud-chewing domestic bovines that till the fields, both release methane. Termites are another source. Is their habitat expanding with destruction of tropical forests?

4.15
Comparison between increases in methane and global population. CH_4 concentrations in ice are represented by green and yellow bars; atmospheric measurements, by gray squares.
4.16
Working the rice fields of Sri Lanka.

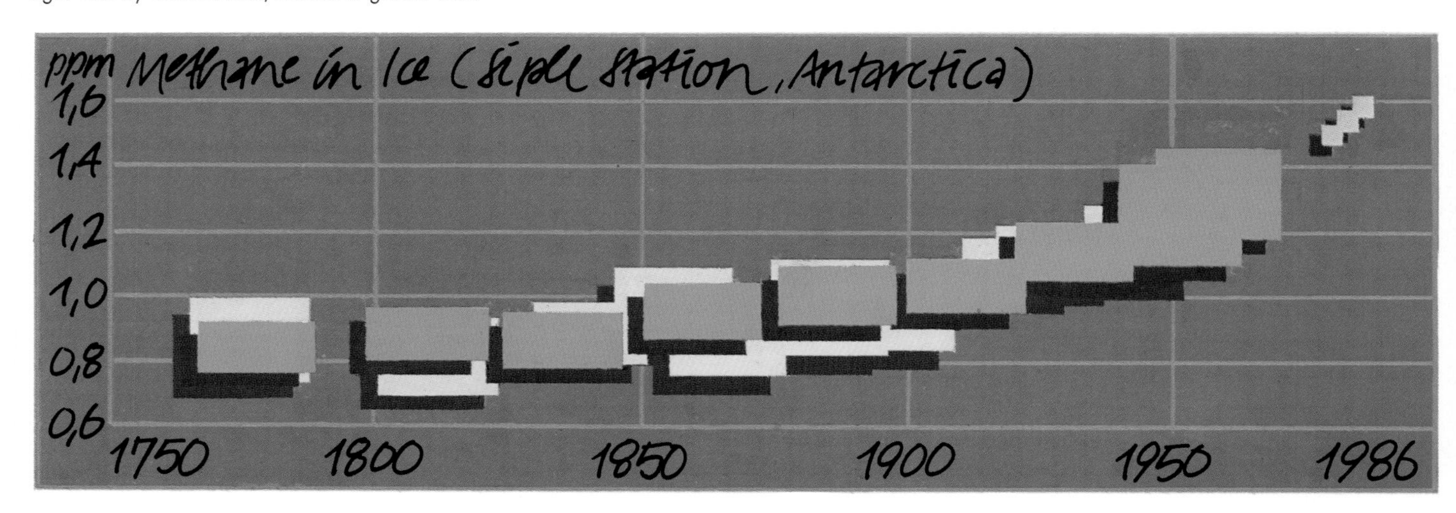

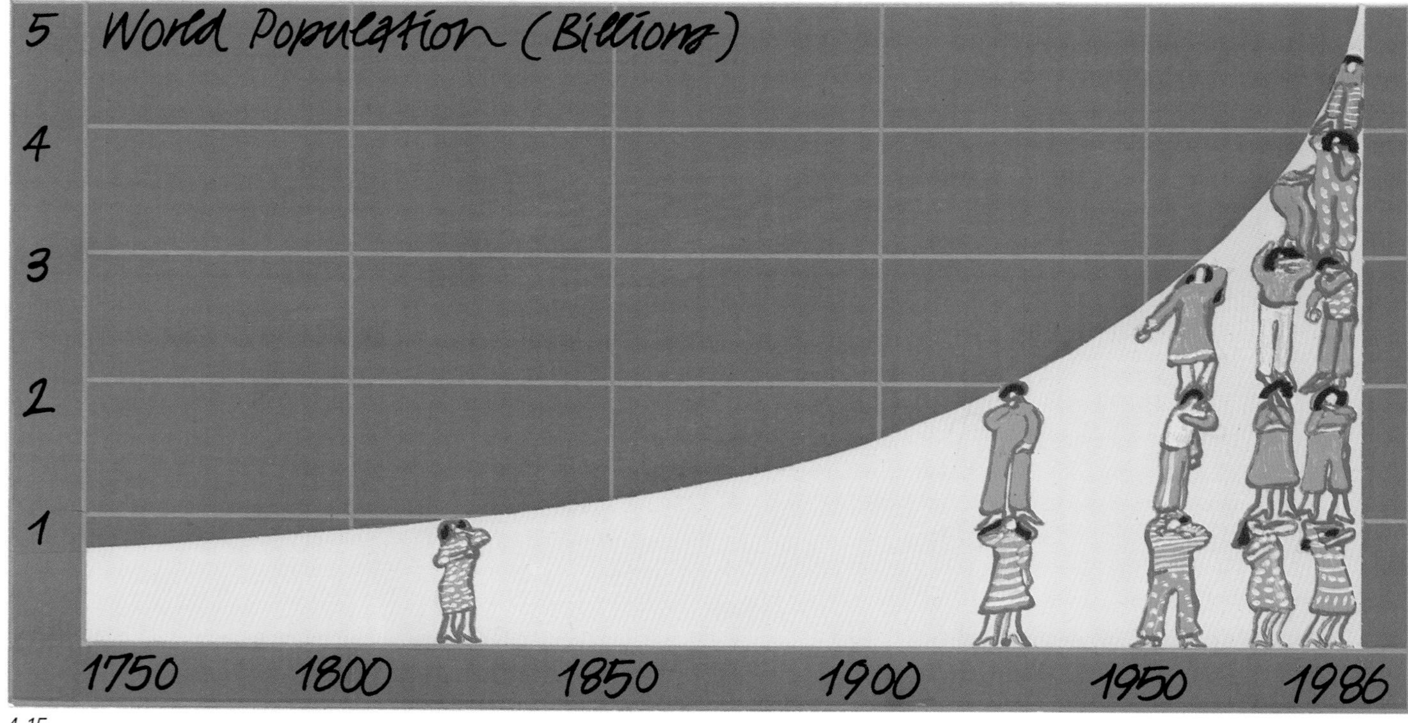

4.15

4.16

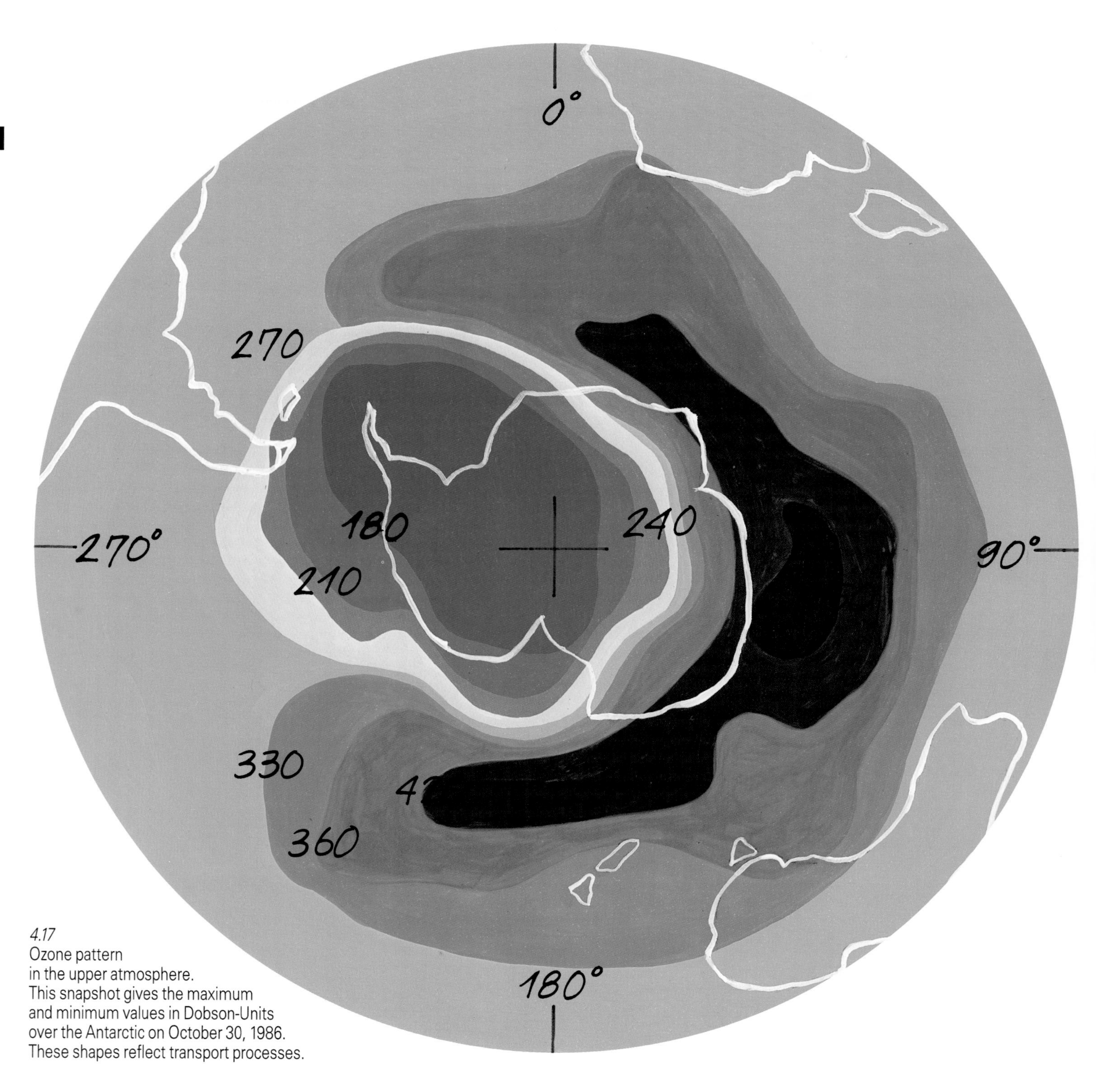

4.17
Ozone pattern
in the upper atmosphere.
This snapshot gives the maximum
and minimum values in Dobson-Units
over the Antarctic on October 30, 1986.
These shapes reflect transport processes.

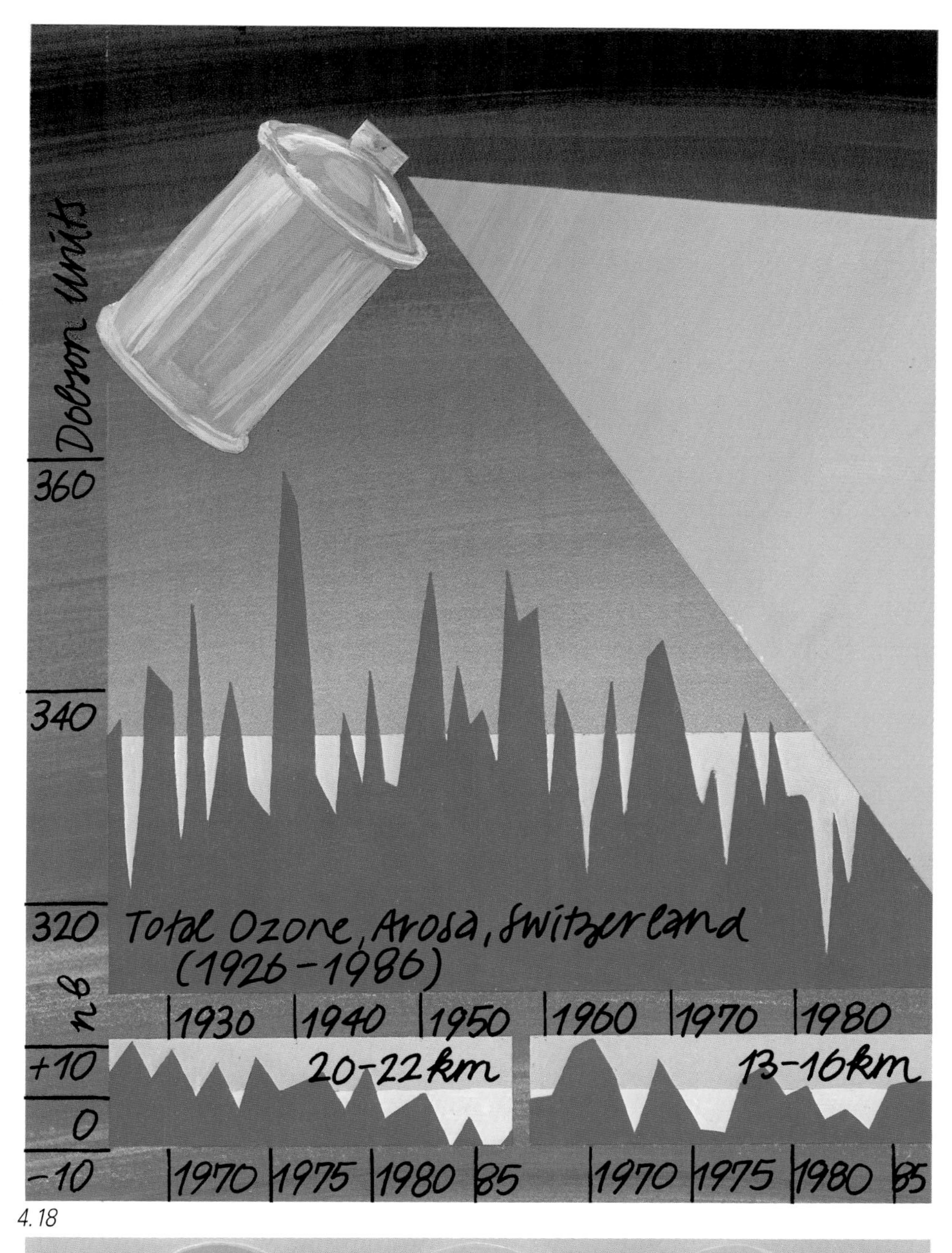

4.18

4.19

Ozone

The discovery of the ozone hole over the Antarctic during Southern Hemisphere Spring opened a long-neglected environmental can of worms. The thin ozone layer in the stratosphere is our protective veil against dangerous ultraviolet rays. On the other hand, ozone in the troposphere is a pollutant and contributes to the greenhouse effect. Human activity disturbs the balance in two ways. With the help of sunlight, increasing amounts of ozone are produced in the lower atmosphere from industrial, automobile and agricultural emissions. Secondly, CFC gases, chlorofluorocarbons, rise higher into the stratosphere. There, via reactions with natural ozone catalyzed by solar radiation, they deplete the ozone layer. CFC's are released from refrigerators, air conditioners, foam rubber, and, above all, from aerosol spray cans. At international conferences scientists and policymakers are intensely bargaining for measures to stop worldwide production before the end of the century. Some industrial countries, including Switzerland, are introducing laws and technology to shorten the deadline. In order to follow suit, the developing countries will need money and the transfer of technology. Even these measures, however, will not provide a quick fix from short-sighted progress. The path of CFC's from the last spray can to the ozone layer will take 10 years. Then they will continue depleting ozone for another hundred years.

4.18
Swiss scientists have measured the longest ozone data set in the world. Ozone values have decreased in the stratosphere and increased in the troposphere. Both trends spell trouble for humans.

4.19
Three snapshots of ozone patterns over the Antarctic in October. Blue, for the smallest and dark green, for the highest values.

4.20
Over: We live in the communication age. International debate can defuse global conflict, or inhibit responsive global action. It is easier to talk than act.

4

4.30

O3

4

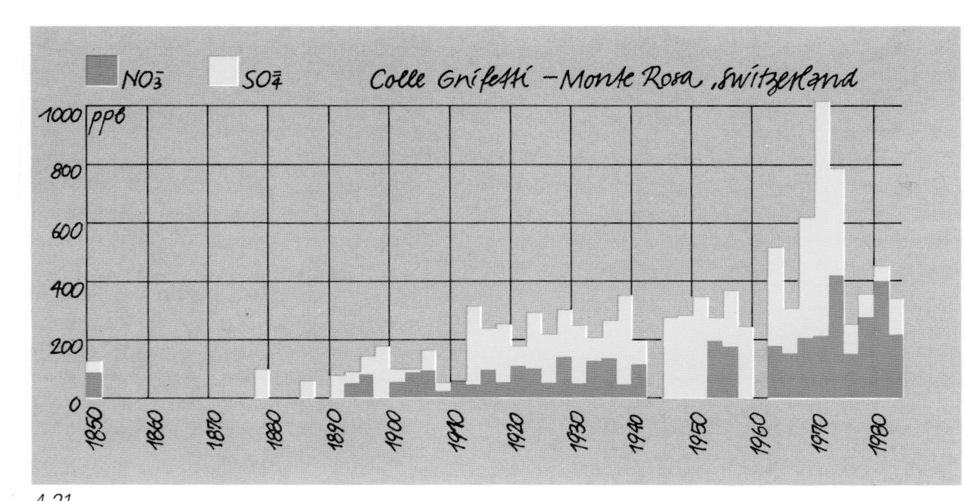

4.21

How sensitive are Humans?

Systematic monitoring of air quality only begins when it literally stinks to high heavens. We have barely more than a decade of worldwide, continuous data. Prior atmospheric states must be reconstructed from the clues in natural archives in order to define human disturbance.

4.21
Bars show a rise of sulfate and nitrate in the atmosphere of central Europe archived in the ice on Monte Rosa, Switzerland.

4.22
Looking down southward from Monte Rosa, across the sea of fog filling the Po Plain, Italy.

4.22

Air that sickens

Polluted city air is a complex mixture. Some pollutants are released directly into the air from various sources; other pollutants form by photochemical reactions. Air quality standards, safe human tolerance levels and the effects of pollutants are subjects of continued debate. The debate heats up whenever atmospheric inversions drive irritants to the limit. Who then turns what off?

4.23
A selection of air-borne components produced by an industrial city, which can affect our health at elevated concentrations.

Cd: Cadmium. Released by waste incineration plants and fired heating. Long term exposure damages kidneys, weakens bones.
Cl_2: Chlorine. From chemical industries. Forms hydrochloric acid. Irritates mucous membranes.
CO: Carbon monoxide. From traffic, heating and steel mills. Weakens the heart.
F: Fluorine. From foundries. High concentrations adversely affect teeth.
Hg: Mercury. From coal and oil heating, foundries. Affects the nervous system, causes trembling.
H_2S: Hydrogen sulfide gas. From refineries, sewage treatment and pulp paper industries. Causes vomiting, irritates eyes.
Mn: Manganese. From the steel industry and thermal power stations. Possible contribution to Parkinson's disease.
Pb: Lead. From automobiles and foundries. Leads to brain damage, high blood pressure.
HNO_3 and H_2SO_4: Nitric and sulfuric acids. Main acid component in rain. Cause lung problems.
NO_2: Nitrous oxide. Sunlight catalyzes its formation from auto and heating emmissions. Causes ozone formation. Causes bronchitis.
O_3: Ozone. Formed by photochemical reaction of sunlight with nitrous oxides and hydrocarbons. Irritant for eyes, promotes asthma.

4.23

4 ...and our environment?

The condition of our forests has rustled the media in recent years. Forests seem to be expressing our bad conscience with respect to nature. They unabashedly expose our mismanagement and leave us completely frustrated by our lack of knowledge of critical feedbacks in the natural ecosystem. The following example from the Swiss midlands inspires reflection. We determine the health of a large number of pine and fir trees from the same area by studying small cores of wood, not only by visual inspection. The form of the annual rings tells us the tree's growth history. We compare reductions in growth thickness with the atmospheric acidity and supply of moisture for the same periods. We can only see a distinct relationship between drought and growth during the hot and dry 1940's. Whereas the fir trees seem to rebound and are little affected by increasing air pollution, the pine trees take a turn for the worse. Does this prove that pines were critically weakened by extreme climate, left susceptible to diseases and finally fell victim to pollution stress? In reality, we don't know the answers and are only beginning to grasp the network of interactions. Progress depends on strengthened interdisciplinary research, not on advocating a single guilty factor.

4.24
Upper: the record of air pollution as measured in ice cores from Monte Rosa. Lower: the record of precipitation during the period of tree growth, expressed as a deficit of moisture with respect to the long-term, local baseline of 450mm per year. The two graphs in the middle trace the development of pine and fir trees during 1910–1980.

4.25
An example of the idealized image of our mountain forest as here in the Austrian Alps – and an inset showing how far destruction must progress before we take adequate notice.

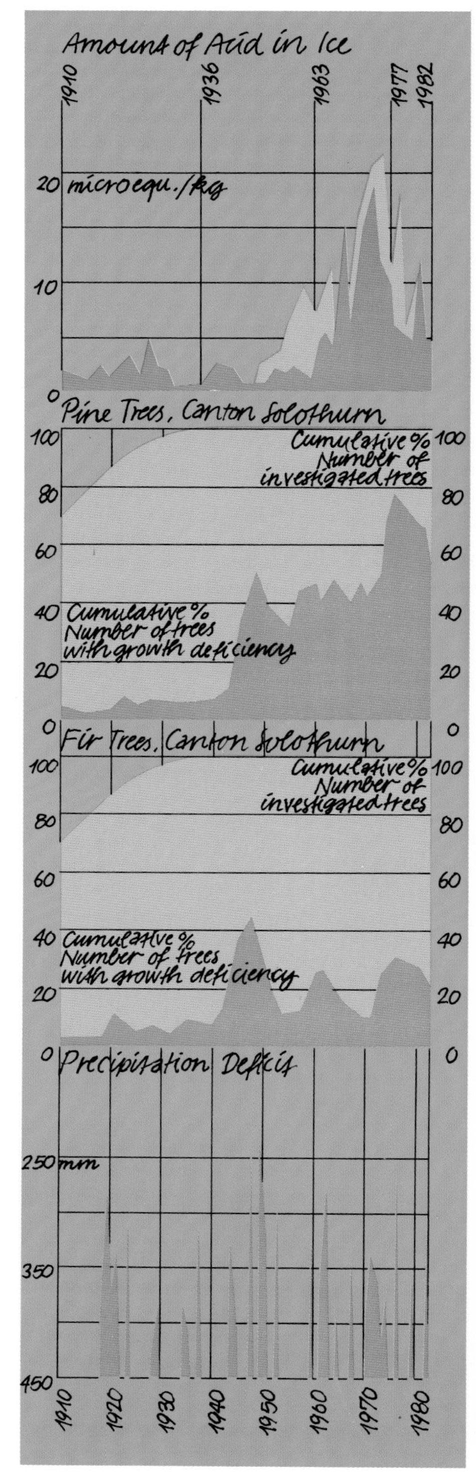

4.24

4.25

4

More people need more to eat. If traditional agricultural methods are improved; if more land is cultivated; if land fertility is maintained; if excess production does not rot at customs borders; if food stuffs can be transported to where they are needed; if, if ...and if... then, the world can support many more children. If climate doesn't double-cross progress. The natural vegetation cover and soil conditions are primarily dependent on climatic parameters. People tilt the scales. The regenerative capacity of meagre soils is limited in most regions of the world. Improper agricultural methods irreversibly deplete nutrient budgets and ecological balances that developed over thousands of years. Wind and water erosion accelerate and complete the sad process, which begins, not always innocently, by deforestation, overgrazing and depletion of soil. When we scan the vistas of the Mediterranean coastline, we are constantly reminded that we are paying a high price for the destructive heritage of our Greek and Roman forefathers. Barren, rocky karst landscapes are left over from devouring lush oak forests. And what, despite our knowledge, is now happening in sensitive areas south of the Sahara, or in many developing countries? On an optimistic note, Man can protect and construct. Even after whole mountain slopes have become destabilized by deforestation, soil loss can be minimized by terracing. Man is a lonely warrior in a landscape devastated by erosion. For a successful communal effort at landscape control, people require a functioning social, economic and political structure.

4.26 and 4.27
Terraced fields in Nepal as an example of successful landscape control.

4.28 and 4.29
An eroded landscape. The surface is prey to hefty rainfall; fertile soil cover is quickly stripped and washed away. An example from Ewaso N'giro, north of Mt. Kenya.

4.26

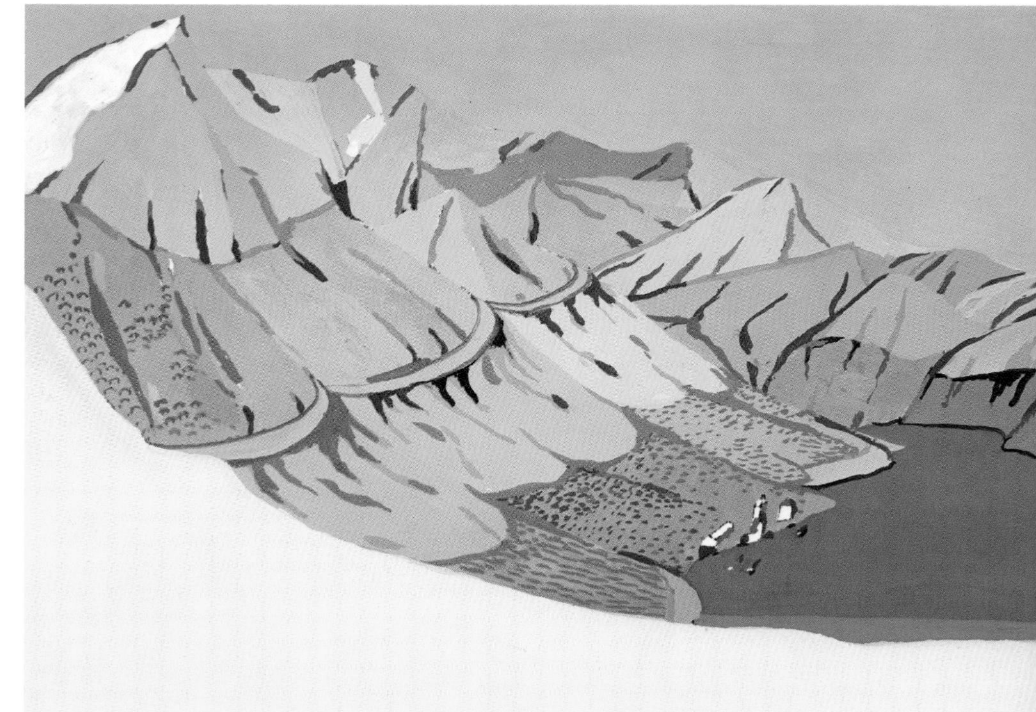

4.28

4.27

4.29

4

Many countries are just in the process of creating new functional structures to promote sustainable development and need help. Industrial countries often lose credibility by providing their aid packages with short-sighted goals in mind. Rain forests are a glaring example. First, financial aid was granted to clear forests for agriculture, although the soils were often inadequate. Then the same countries were promised additional aid for reforestation to dampen the increase of atmospheric carbon dioxide.

4.30
External aid is only of use when it fits harmoniously into the fabric of the local system. Superiority alone can be destructive.

4.31
Does burning the tropical forest provide the best foundation for successful agriculture?

4.30

4.31

The 1988 Toronto World Conference on the Changing Atmophere made it clear that our only chance to limit the impact of our uncontrolled experiment with the atmosphere is to reduce North-South inequalities. Debt reduction and technology transfer have not yet advanced much beyond media headlines. Action is demanded rather than delay tactics, couched in the rhetoric, which claims incomplete knowledge. The inertia of governments and societies does not allow us the luxury of wasting time to develop efficient strategies to counter global change.

4.32
The societal puzzle for confronting global change.

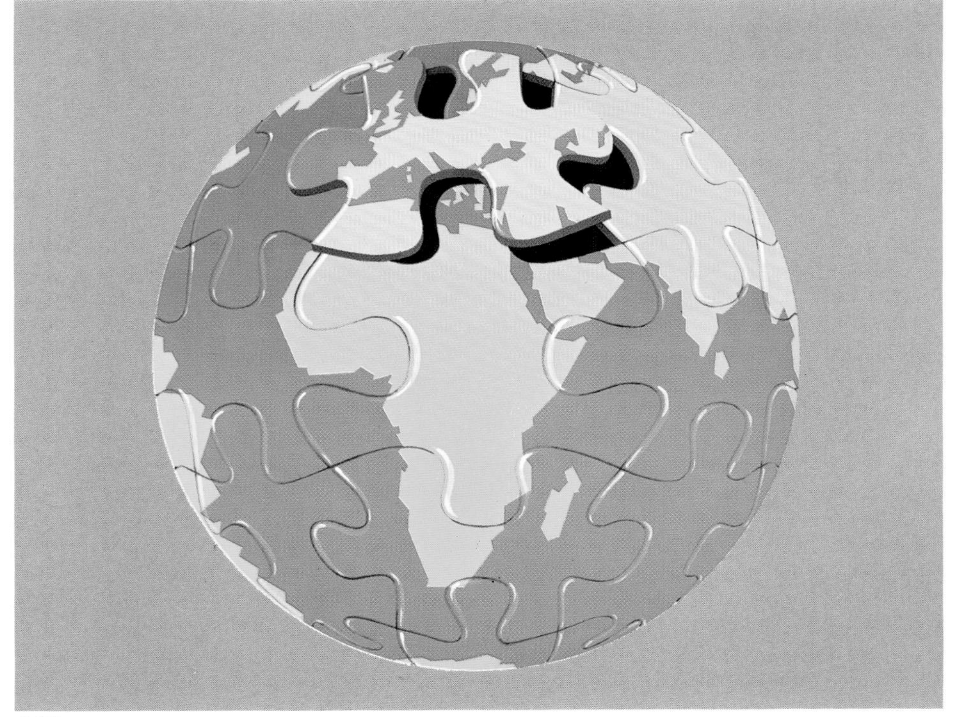

4.32

4 Man and climate

To the scientific community represented by IPCC, it is clear that emissions, resulting from human activities are substantially increasing the atmospheric concentration of the greenhouse gases carbon dioxide, methane, chloro-fluorocarbons – CFC's – and nitrous oxide. These increases will enhance the natural greenhouse effect, resulting in an additional average warming of the Earth's surface. The main greenhouse gas, water vapour, will increase as a possible feedback response to global warming, further enhancing the effect.

Carbon dioxide presently accounts for more than half of the enhanced greenhouse effect, and will probably continue to do so during the next decades. Since the industrial revolution, combustion of fossil fuels and deforestation have led to a 26% increase in the atmospheric concentration of CO_2. The present operation of the carbon cycle is such that about half of the emitted CO_2 stays in the atmosphere; the remainder is taken up by the ocean and the biosphere. It takes 50–200 years for the atmospheric CO_2 level to adjust to a new equilibrium, mainly because of the slow exchange of carbon between surface and deep waters in the ocean. Deforestation reduces CO_2 uptake by the biosphere; thus it represents a source for CO_2. Reforestation to remove 5–10% of the CO_2 emitted by fossil fuel burning would require seeding a surface area 100 times as large as Switzerland by the year 2030. In the meantime, forests will continue natural cycles of maturation and decline, particularly in a climate to which they are increasingly poorly adapted.

Atmospheric methane concentrations have more than doubled during the last 150 years. Rice and cattle production, biomass burning and the ventilation of natural gas are the main reasons. The present growth rate is about 1% annually. In a warmer climate-particularly in tundra environments-dead plant matter will decay faster, leading to an increase in natural methane sources. Though the atmospheric lifetime of methane is only 20 years, it is 20–30 times more efficient than carbon dioxide at trapping heat and thus creates a major climate feedback.

Nitrous oxide has increased by about 8% since pre-industrial times. Currently, we are unable to specify all of the sources of N_2O, but agriculture plays a part. The main natural source is microbial denitrification in soils. Humans enhance this source through the use of nitrogen fertilizers. The conversion of forest areas into farmland, therefore, will contribute also to the increase in N_2O emissions. Nitrous oxide also has an important destructive effect on stratospheric ozone concentrations.

The CFC's were not present in the atmosphere until their industrial production prior to WW II. Mainly used as aerosol propellants, solvents and foam blowing agents, CFC concentrations have increased dramatically since 1960. In the last decade they contributed more than 20% to the enhanced greenhouse effect. The CFC's have attracted public interest, because of their destructive impact on the stratospheric ozone layer, which allows more cancer-causing UV-B radiation to penetrate the Earth's atmosphere. Ozone depletion in the stratosphere should also have another, more welcome effect: cooling the Earth surface in a negative feedback response to the enhanced greenhouse effect. It is absurd, however, to compare a possible benefit with the increased threat to life that stratospheric ozone decreases represent.

Although its climatic effect cannot yet be accurately quantified, ozone at ground-level is a dangerous air pollutant, particulary in the Northern Hemisphere. It was not until about 20 years ago, when air literally stunk to high heaven, that the battle to reduce air pollution began. Today, despite remarkable progress in clearing up some forms of air pollution, the World Health Organization estimates that 70% of the global urban population breathes air that contains unhealthy levels of pollutants, at least part of the time. Fortunately, any further efforts at pollution control are likely to result in lowered greenhouse gas emissions, as well.

Climate research

5

Climate research sits in a niche between the exact and descriptive natural sciences. The challenge of global change requires a cross-over among disciplines. Scientists continue searching for reliable scientific evidence,information on potential impacts and reasonable countermeasures. We are pressed for time. From what we know today, it is clear that we cannot assume that climate change will be gradual and leave us adequate time to adapt. Changes can be unpredictably abrupt, with sudden steps and relapses of varied amplitude and duration. In order to meet this global challenge, we must seize the opportunity to strengthen the global web of cooperative, goal-oriented research on both national and international levels.

5 Past climate research

Ice Ages are a mystery with clues that link global climate and local weather. The theory of the Ice Ages has its roots in Switzerland. In the 18th Century, an alpine goat herder in Grindelwald and a ram hunter on the other side of the mountain in Valais, independently proposed a theory, which became the butt of major dispute among scientific naturalists. They claimed evidence that glaciers had once expanded and filled the valleys. Louis Agassiz from Neuchâtel finally deluged the opposition, and anchored the Ice Age concept in the international scientific community. These ideas, together with the equations of Milutin Milankovich, still form the basis of our understanding of the variability of climate

Systematic archiving of instrumental weather data began soon after the invention of the barometer and thermometer. The first pressure-pattern map was published in 1813, inspired by strange sandstorms in the Alps. The invention of the telegraph in 1843 led to immediate exchanges of weather reports, well beyond national boundaries.

5.1
Louis Agassiz, a pioneer of the Theory of Ice Ages, lead a research program from 1841–1845 on the Aar Glacier in the Bernese Alps. He and his colleagues utilized the shelter of a giant erratic boulder on the medial moraine that became famous as the Hôtel des Neuchâtelois.

5.2
Changes in glacier length are an expression of climate fluctuations. Glaciers have been systematically surveyed in Switzerland for years, because of their importance for hydroelectric power. The black band denotes the percent of glacier tongues that neither advanced- in light blue- nor retreated – in gray.

5.3
Early instruments for meteorological measurements at 2500m in the high altitude Säntis Observatory, Switzerland.

5.4
Heinrich Wild, front row, right, was director of the Bernese Astronomical Observatory and one of the founders of the World Meteorological Organization, WMO.

5.1

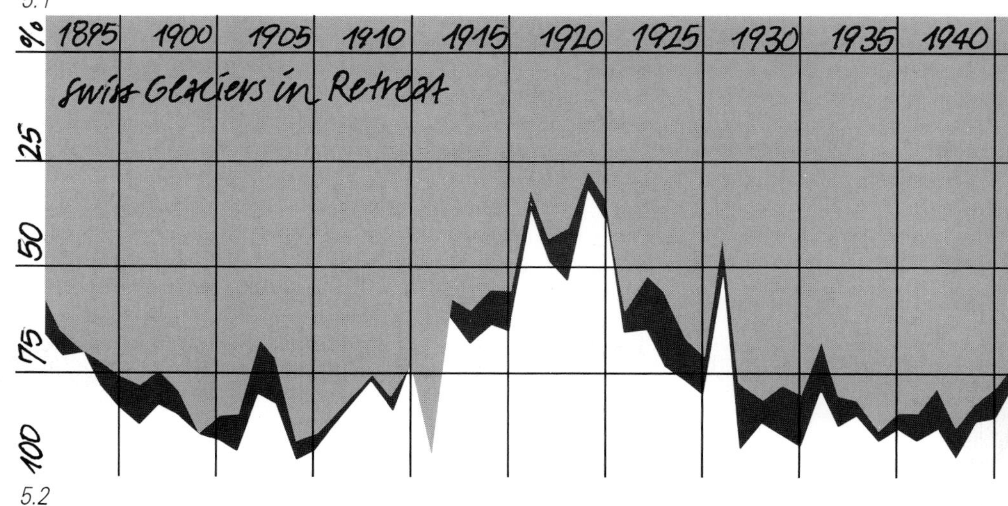

5.2

5.3

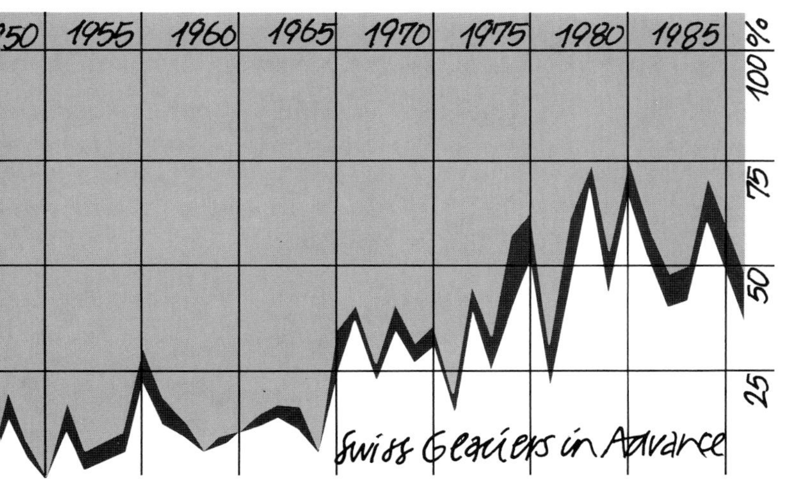
1955
1960
1965
1970
1975
1980
1985
100%
75
50
25
Swiss Glaciers in Advance

5.4

5

In the late 19th Century, meteorological services sprouted in many countries. The Swiss Meteorological Institute evolved from a commission of the Swiss Academy of Sciences during a time of increasing demand for agricultural weather forecasts. Continuous series of meteorological measurements formed the scientific basis for stastistical evaluation of the weather's moods. We are still working on homogenizing these data for worldwide comparisons. Many people require direct evidence of rising global temperature before they accept the role of human impact on our climate system. Unfortunately, most meteorological networks only began measuring at the same time that industry began enhanced emissions of greenhouse gases. When trying to evaluate the global temperature trends of the last 130 years, we are haunted by the fundamental questions of what is natural variability and what is human impact?

5.5
The Hospice on the Great St. Bernhard Pass between Switzerland and Italy. It was the first mountain weather station which continuously recorded temperature.

5.6
Temperature record from Great St. Bernhard Pass reported as deviations from the mean. The data illustrate extreme cold phases at the close of the Little Ice Age, which led to a wave of emigration. They also show the marked warming trend in the 20th Century.

5.5

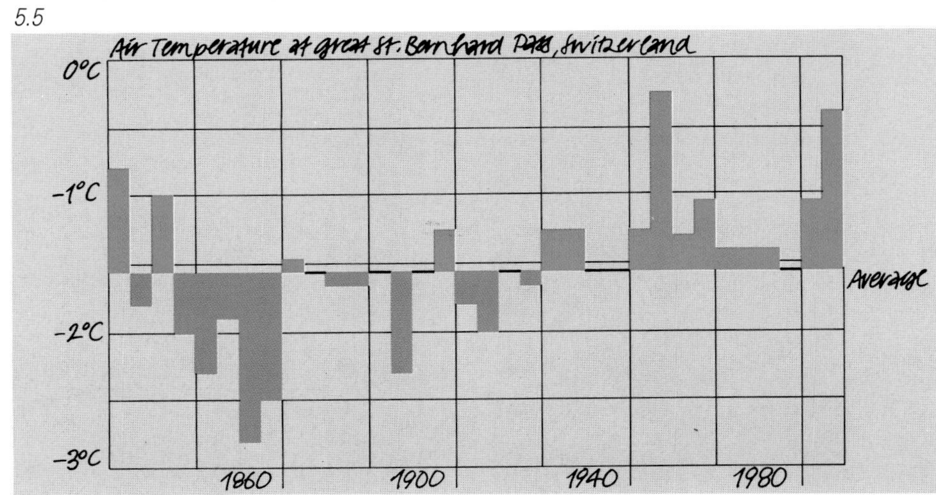

5.6

Present climate research
Air pollution

Human impact on climate and environment is first felt on the local or regional scale. Industrial, automobile and household emissions of pollutants increased markedly after World War II. The consequences were first buried in the euphoria of economic growth. Water pollution, acid rain and forest damage grow into the headlines of industrial areas. Air is an interesting subject of research. Air pollution stops for no border. Complex regional transport is simulated with the help of laboratory experiments in order to find a scientific basis to define political measures. round level air currents are traced by passing smoke over model landscapes. By adding artifical barriers representing cities or industrial centers, modellers predict the effects of human interference. Major interest focuses on smog and the conditions responsible for increases of tropospheric ozone.

5.7
Experimental results are simplified to a model view of processes near the planetary boundary layer, at the bottom of the atmosphere. In this region we live and breath.

5.8
A scientist watches the behavior of air currents in a laboratory experiment.

5.9
Smog formation in the Grindelwald valley. Automobile exhausts react with sunlight to produce ozone. The long-term trend is upward.

5.7

5.8

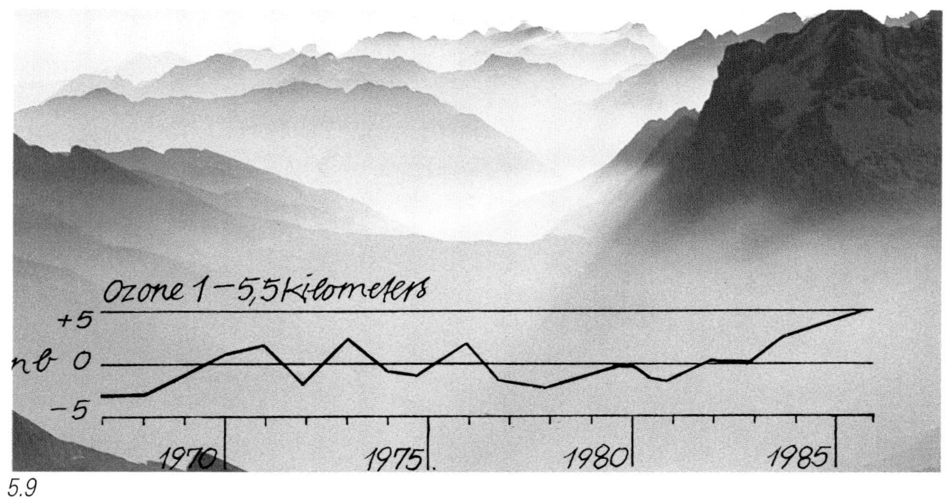

5.9

5 Global dust transport

A vineyard between glaciers. What does this mean for climate research? We view the world from space. We apply the fastest computers to crunch an avalanche of current climate data for the quickest publishable results. But this knowledge of climate only strokes the skin if we do not probe lessons of the past. How else can we say whether today's observations are chance, or follow logically from processes beginning long before our present technical snapshots? Satellites allow us to trace large-scale dust transport. Fossil soils and glaciers archive the evidence of these movements. Saharan dust often sprinkles the Alps. Dust layered into Monte Rosa glacier ice adds more pieces to the puzzle which defines air-pressure patterns over the last 500 years. For a million-year record of air movements, we have thick layers of dust called *loess deposits.* Loess regions of the globe lie along deserts rims, or in arid regions along the fronts of ice sheets. Winds lift dust from moraines and glacial outwash deposits

5.10

5.11

and carry it thousands of kilometers. In China, loess from inner Mongolian deserts built quilts, hundreds of meters thick, which smoothed the precursor landscape into rolling hills and dales. Loess provides a fertile soil for agriculture.

5.10
At 5500 m high in the Russian Pamir, a tilted block of ice displays dust layers at an angle. The dust blew from central Asian deserts.

5.11
A vineyard in Austria growing on gentle loess hills. Nearby, Czechoslovakian scientists study 130 000 years of climate history in a thick sequence of loess layers.

5.12
Several layers of yellow Saharan dust in ice cliffs of the Monte Rosa Glacier

5.12

5 The carbon dioxide paradox

Why doesn't the ocean swallow more CO_2? Immense quantities of CO_2 are stored and transfered between the giant reservoirs of the oceans and biosphere. Compared with these, the amounts humans add are trivial. Still, atmospheric CO_2 increases. Why? The chemistry and biology of the surface ocean limit CO_2 uptake. Further transfer into deeper water layers is slow. Thus ocean circulation excercises an important control on climate. Researchers are onto the trail of these transport mechanisms by dating the shells of microfossils. Foraminifera near the surface live in waters of a different water age than those living on the seafloor at the same time. These different ages are stored in the carbon of their shells, which are buried together in the sediment. Time series of these age differences show changes in rates of ocean mixing in the past.

5.13 and 5.14
Big machines are required to measure the ages of very tiny life forms.

5.15
Carbon reservoirs and their fluxes.

5.16
Differences in shell ages between Foraminifera from the surface and from the seafloor are a measure of how long it takes the ocean to mix.

5.17
Age differences between surface and 3 km depth. Red dots where surface water sinks.

5.13

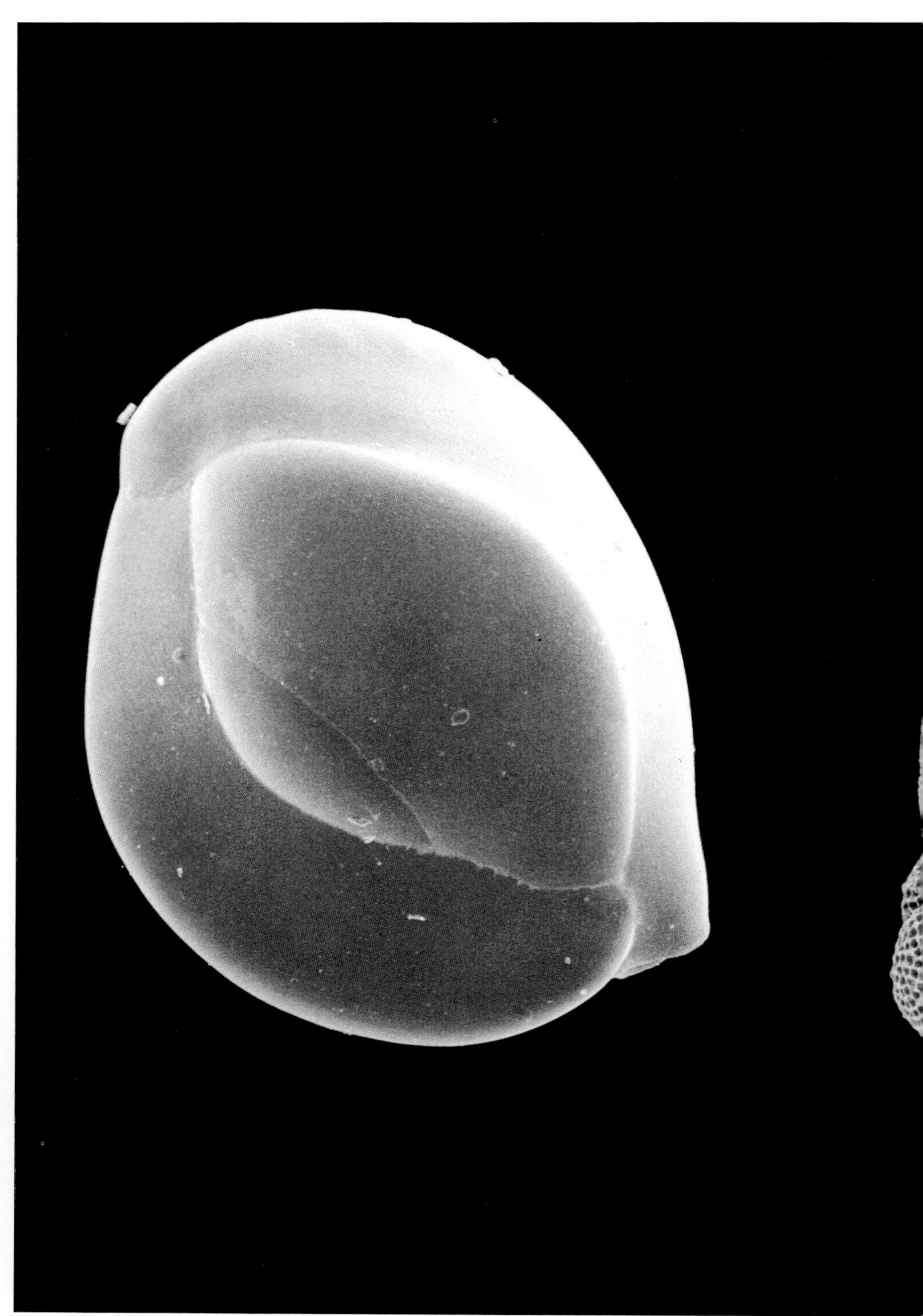

5.14

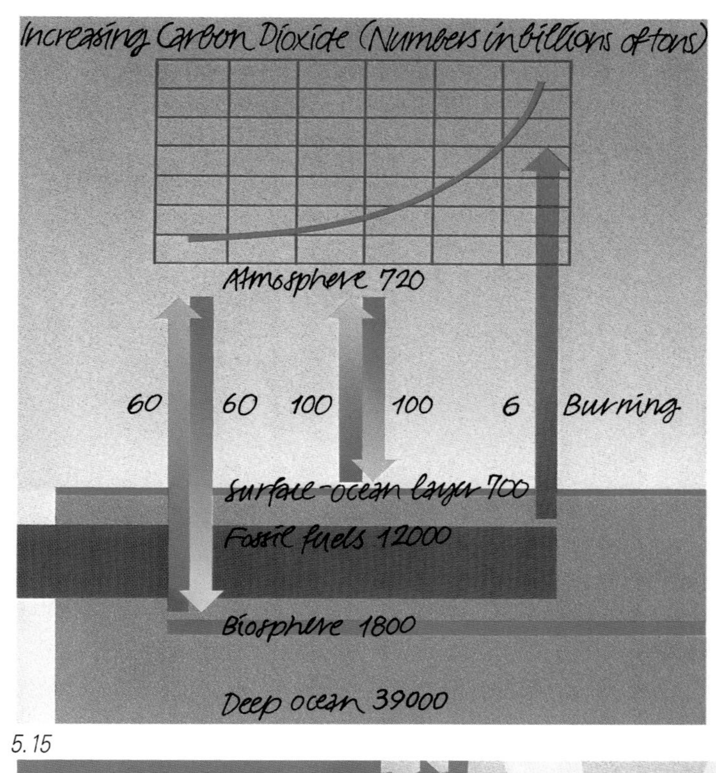

5.15

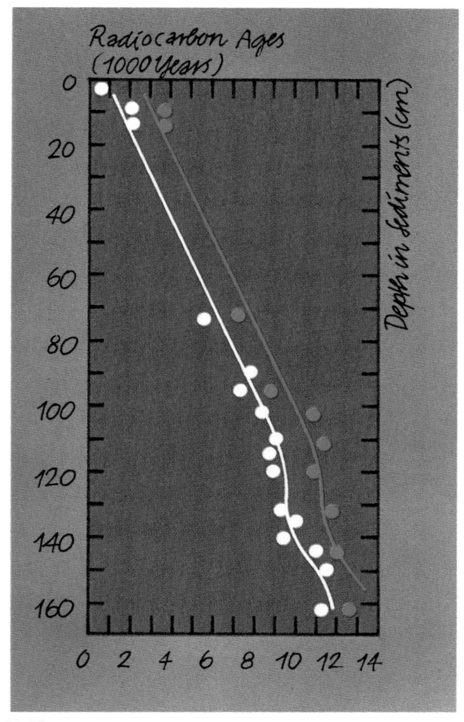

5.16

Age Distribution of Ocean Water at 3 km Depth (Years)

1500
1750
250
1750
1250
1000
500
750
1000 1250 1500
500
500
750
500

5.17

5

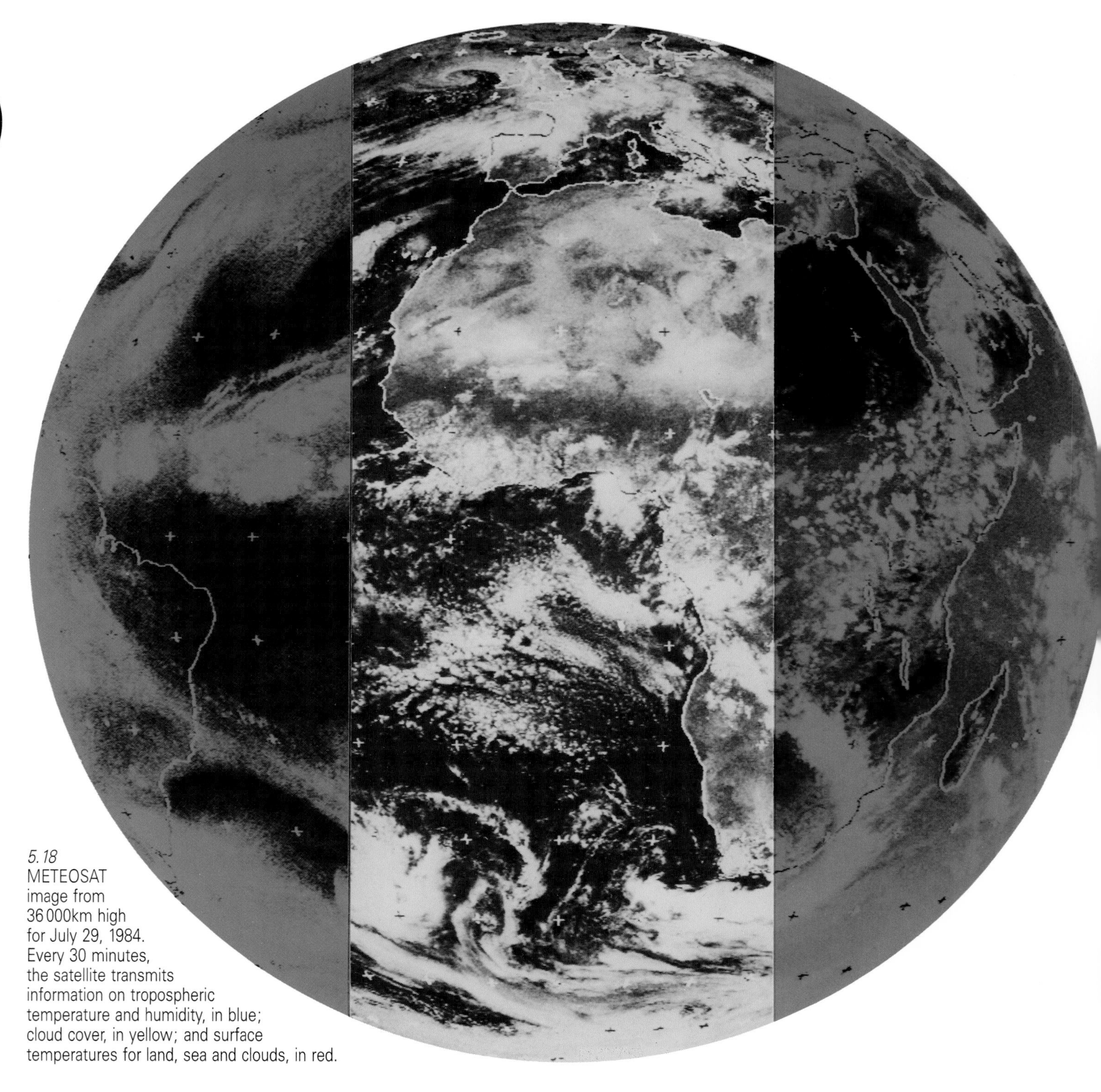

5.18
METEOSAT
image from
36 000km high
for July 29, 1984.
Every 30 minutes,
the satellite transmits
information on tropospheric
temperature and humidity, in blue;
cloud cover, in yellow; and surface
temperatures for land, sea and clouds, in red.

5.19

Mission to Earth

Single data points are inadequate for understanding climate or the complex processes in the atmosphere. We require global data coverage to test whether computer simulations of climate are anywhere close to reality. Many skeptics of the race to the moon now concede that it brought back a new view of the Earth. The Earth is now spanned by a sophisticated network of standardized weather monitors. The importance of satellites lies in their ability, not only to directly collect data on a global scale, but also to query, record and transmit data from ships, weather buoys, or remote land stations to the international weather data centers.

5.19
A worldwide weather-recording network is necessary for understanding the processes that control climate.

5.20
Time lapse movements of cloud cover at 3-hour intervals. Cloud dynamics, as here over central Europe, are familiar to us from TV weather reports.

5.21
Sea surface temperature of the Atlantic Ocean recorded from a single satellite sweep.

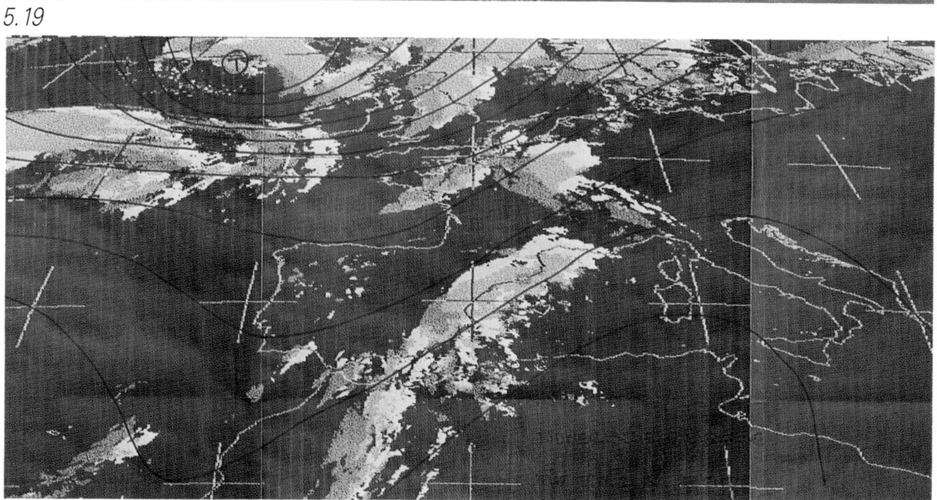

5.20

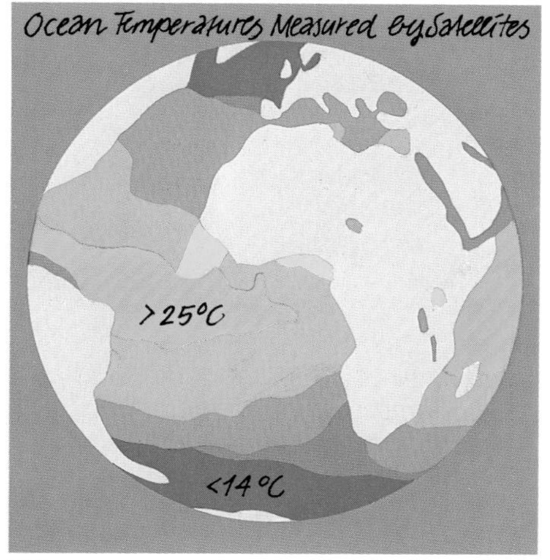

5.21

5

The global web

The complexities of global change are beyond the scope of single researchers. Scientists have begun cooperating in a matrix of global programs with the help of both national and international agencies.

World Climate Research Program

The WCRP is one of four parts of the World Climate Program founded in 1979 at the First World Climate Conference in Rome. It encourages national programs with similar goals. The objectives of WCRP encompass three time scales that affect society: monthly and seasonal weather forecasting and the prediction of decadal climate changes. *Can we extend weather forecasting?* Atmospheric circulation is, by nature, unstable, hence unpredictable for periods longer than 10 days. Under special conditions, pressure cells can remain stationary for weeks. By probability analyses,one hopes to extend the forecast period up to a month. *Can we predict shifts in seasonal behavior?* Thanks to the catastrophic effects of the 82–83 El Niño events, we now know more about atmosphere-ocean coupling and changes in average weather from year to year. TOGA, the Tropical Ocean and Global Atmosphere Study, targets El Niño/ Southern Oscillation phenomenon and the variations in the monsoon regime. TOGA aims to test whether these systems are predictable on monthly and annual time scales. *Where does it dry,where does it flood?* We cannot predict how climate belts will shift or how climate readjusts to external forcing without knowing how the ocean responds to the same stresses. The World Ocean Circulation Experiment, WOCE, seeks to improve the treatment of the ocean as a moving media which is dynamically coupled with the atmosphere. It is hoped that such models will identify regions endangered by decadal droughts.

5.22 and 5.23

The magic of powder snow, dream of millions of Europeans. Can we intrust our future to a calendar? Remember the mild winters of 1988–90. Note the1983–84 shift in snow cover with respect to a 40-year monthly average. Further shifts seem likely. For long-range vacation plans call WMO.

5.22

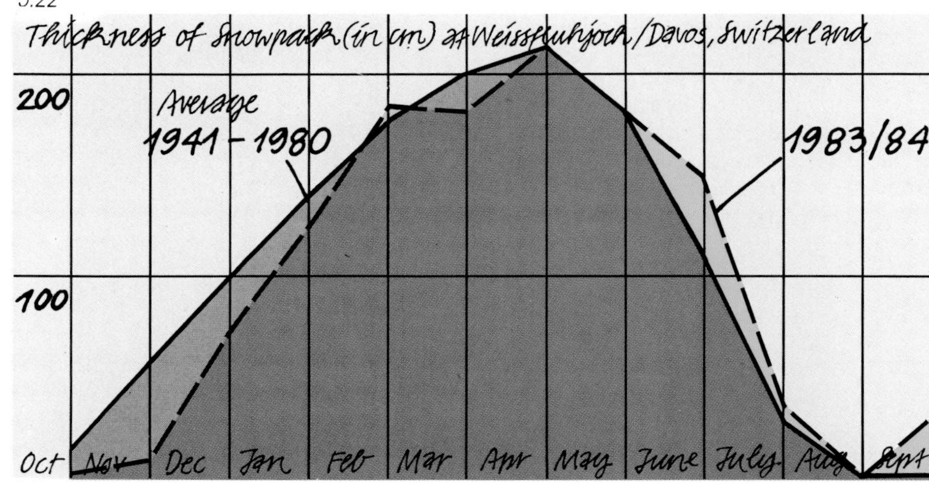

5.23

International Geosphere-Biosphere Program

In September 1986, the International Council of Scientific Unions, ICSU, agreed to establish the International Geosphere-Biosphere Program: A study of Global Change.The objective of IGBP is to describe and understand the interactive physical, chemical and biological processes that regulate the total Earth system and the manner in which they are influenced by human activities. In 1990 the program consists of 10 core projects; two are described below.

Joint Global Ocean Flux Study, JGOFS, is run together with the World Climate Research Program. Scientists want to know how carbon and associated elements move among the major reservoirs in the ocean , biosphere and atmosphere. A central question is the role of the ocean in the uptake of CO_2, and as a buffer of the enhanced greenhouse effect caused by fossil fuel burning.

Past Global Changes, PAGES, aims to reconstruct quantitative information on global changes from natural archives of the past. They will be used to understand large-scale responses of the Earth system to natural and human-induced forcing factors. The research activities are divided into a Stream I to look in detail at Earth history during the past 2000 years, and a Stream II to deal with dynamic feedbacks in Glacial and Interglacial Cycles

5.24 and 5.25

Palms bow to monsoon rains. Prayers are answered for sown fields. Better forecasts of the onset and distribution of rainfall would increase the chance of blessings. Changes in the biosphere are closely related to deficits and surpluses in precipitation. The graph below shows percentage of land area in India receiving either too much or too little rain. IGBP studies couple the physical climate system with biosphere interactions.

5.24

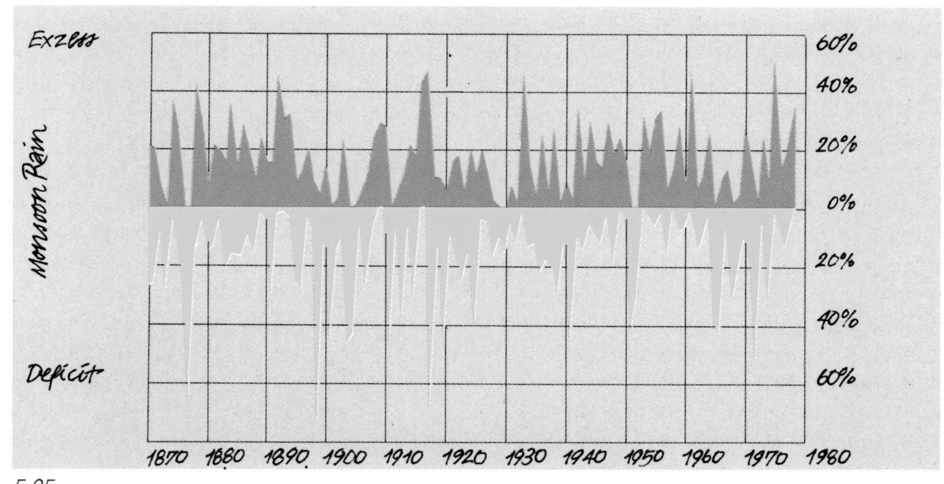

5.25

5 Intergovernmental Panel on Climate Change

In 1988, the World Meteorological Organization and the United Nations Environmental Program jointly invited all nations to participate in a fundamental assessment of climate change. Industrial and developing countries worked as partners from the outset. The report of the experts form the basis of the Second World Climate Conference of 1990. Working groups were divided into: Scientific Assessment, Potential Impacts and Response Strategies.
Working Group 1 compiled all avaliable scientific information on the natural climate system and human interference. They applied the following criteria: what do we know with certainty; what do we calculate with confidence; what is predicted by current models; and what is our best judgment on matters that are more uncertain. *Working Group II* had the more difficult job of predicting impacts of climate change on land, the water cycle and ecosystems. Many predictions cannot be confirmed until they occur for inherent reasons of basic physics or because the state of the future atmosphere has no past analogue. Society's reactions to unexpected changes automatically impact on the findings of *Working Group III*. Theirs was the struggle of formulating realistic response strategies based on energy scenarios. A major dilemma of the climate change issue is that action is needed before proof. IPCC is unlikely to serve a palatable menu for all political tables but is likely to continue cooking.

5.26 and 5.28
Encroachment of the desert sands or widespread flooding: who, when and where is hit?

5.27
Normalized deviations of rainfall in the Sahel Zone. El Niño events are marked by triangles. The correlation with arid years seems obvious. We need better forecasts in order to develop strategies for survival.

5.26

+1
-1
1900 1910 1920 1930 1940 1950 1960 1970 1980

5.27

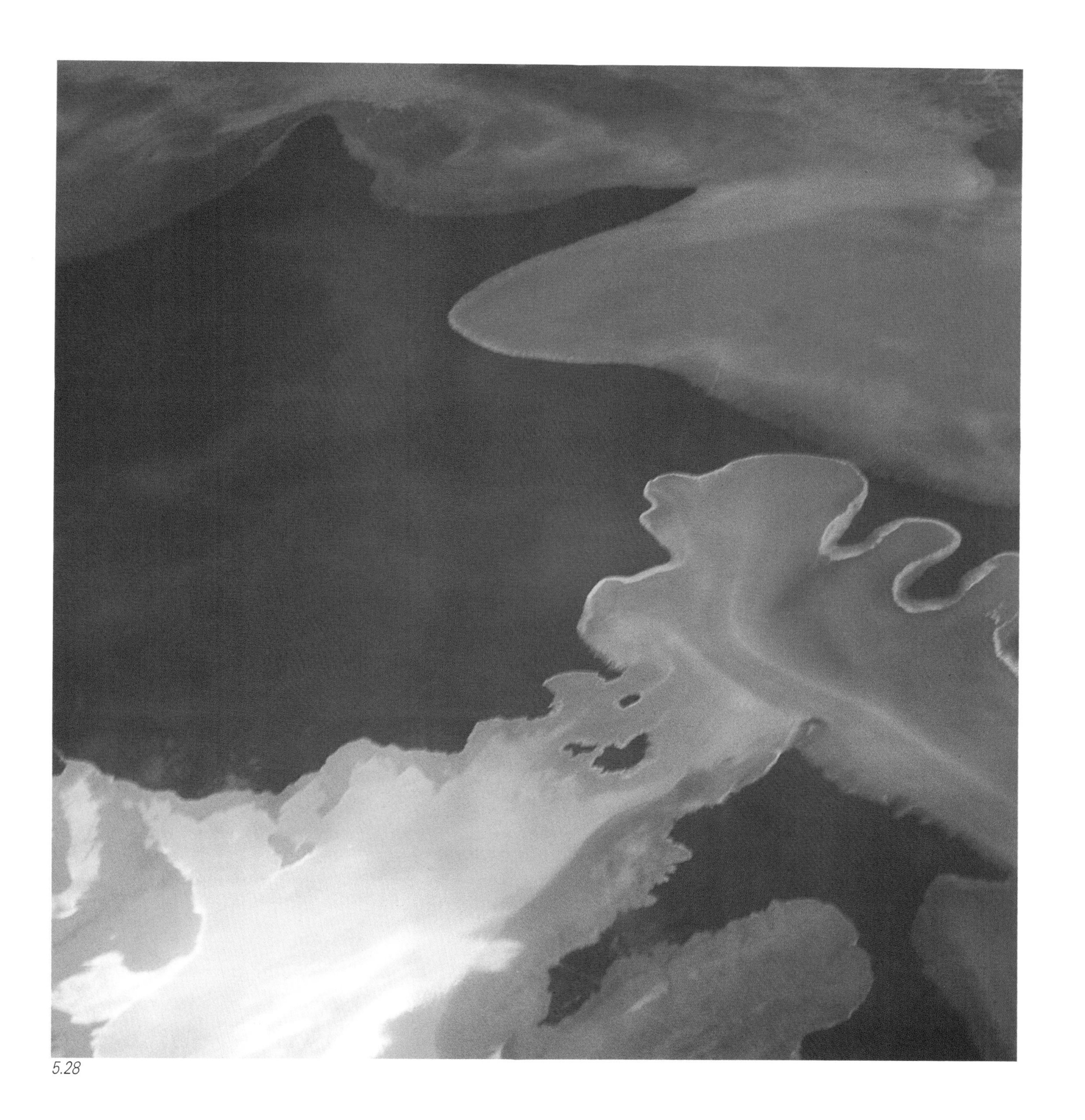

5.28

5

Conference Activities

Can scientific and policymaking conferences solve the problem of global climate change? Such activities have mushroomed in recent years, but when will we see the first steps from discussion to action?

The Second World Climate Conference, SWCC, held in Geneva in late 1990, was the culmination of activities on the world stage over the preceeding two years. In June 1988, the Toronto World Conference on the Changing Atmosphere described the ultimate consequences of uncontrolled anthropogenic air pollution as *second only to a global nuclear war.* In November of the same year, the IPCC began; and in December, the UN General Assembly adopted Resolution 43/53 on the *Potection of global climate for present and future generations of mankind.* At The Hague in March 1989, 24 heads of state agreed to promote new institutional authority for protecting the atmosphere within the framework of the UN. In May 1990, ministers met in Bergen and drafted a declaration on sustainable development. This steam roller was slowed in July 1990 at the Houston summit, which called for more research before policy.

The public turned to Geneva in expectation of action. A six-day scientific session for 500 invited experts considered the IPCC report and future priorities of the World Climate Program. The two-day meeting at ministerial level that followed resulted in a declaration. The meeting was the first small step towards a world climate convention, but no coordinated action to reduce greenhouse gas emissions resulted. Only from a scientific point-of-view did the conference make real progress: the IPCC Scientific Assessment was adopted as a concensus view of our present understanding of the climate system. The public learned that the scientific community is certain of the following:

There is a natural greenhouse effect which already keeps the earth warmer than it otherwise would be. Emissions resulting from human activities are substantially increasing the

5.29

The Jet d'Eau fountain of Geneva balanced the organizers of the Second World Climate Conference.

5.29

atmospheric concentrations of the greenhouse gases: carbon dioxide, methane, chlorofluorocarbons (CFCs) and nitrous oxide. These increases will enhance the greenhouse effect, resulting on average in an additional warming of the Earth's surface. The main greenhouse gas, water vapour, will increase in response to global warming and further enhance it.
With a sound scientific basis, public attention focussed on the 1992 UNCED, the United Nations Conference on Environment and Development, Rio de Janeiro. This conference is another test case for setting political will into action.

5.30
Is UNCED '92 the conference that points the way to sustainable development?

5.30

5 National Climate and Global Change programs: The Swiss example

Many nations have national programs to guide climate-related science and policy, each with its own flavor. Several national programs marry the goals of the World Climate Program and those of the International Geosphere-Biosphere Program. Interactions between the biosphere and the physical climate system are intimately linked. Disciplines that touch on climate and Global Change problems have developed independently. Their approaches to understanding the complexity of natural systems move on different planes of time and space.

Since 1983, scientists have been working under the auspices of the Swiss Academy of Sciences to establish a long-range research program on the issues of climate and Global Change. Switzerland has recently experienced the warmest years on record, three consecutive winters lacking snow and extended periods of valley fog. Sudden, strong storms caused catastrophic floods, record winds damaged forests and droughts in the southern Alps heightend fire danger and water problems. The media promotes a sense of apocalypse. The public and scientific community both ask if the situation is still normal.

Basic research must face new and immediate pressures of long-range forecasts in complex systems. The nagging question is *How quickly must we know what?* ProClim – the National Institute for Climate and Global Change – was inaugurated in 1988, partly in response to calls from the WCRP. ProClim is trying to pull all the players together to build a common offensive towards shared interdisciplinary goals. The institute has four main activities:

Initiating and guiding long-term research projects on climate and Global Change problems; promoting high-quality interdisciplinary research; coordinating Swiss and international research programs; and promoting a dialog among scientist, the public and policy makers.

Research strategy is based on national interests framed in a global context. In the ABC's of ProClim research, A is for improving our understanding of climate processes, modelling and data, B is for reading lessons of the past in climate archives and C is for regional impact studies to evaluate the sensitivity of our ecosystems to climate change.

A-level projects include an intensive analysis of all Swiss weather data, wich has been hamstered for over 100 years. Will it be possible to distill an essence of change? Another project is modeling alpine climate. Climate modelers treat data like toys. The Alps, a dynamic natural laboratory, provide an infinite sandpile of data. Will modelers succeed in weaving the alpine chain into the fabric of global climate models?

B-Level projects include the study of signatures of change in sediment cores from lakes. One focus is rates of natural change at the beginning and end of the Younger Dryas cooling event during the last deglaciation. Can we discover thresholds that trigger unexpected responses in exosystems? Another project is to calibrate ice-core, historical, tree-ring, and sediment-core archives with the instrumental record. Will they come together?

C-Level projects include integration of climate and ecosystem parameters at a series of critical threshold sites in the Alps. Can we identify criteria for their potential responses to specific climate scenarios? Another project deals with the risk perception of regional populations to rapid changes of their environment. The farmer as landscape gardener?

5.31
Landsat image on a cloudless day in spring. The Alps and much of the Jura mountains are under deep snow cover. How much longer?

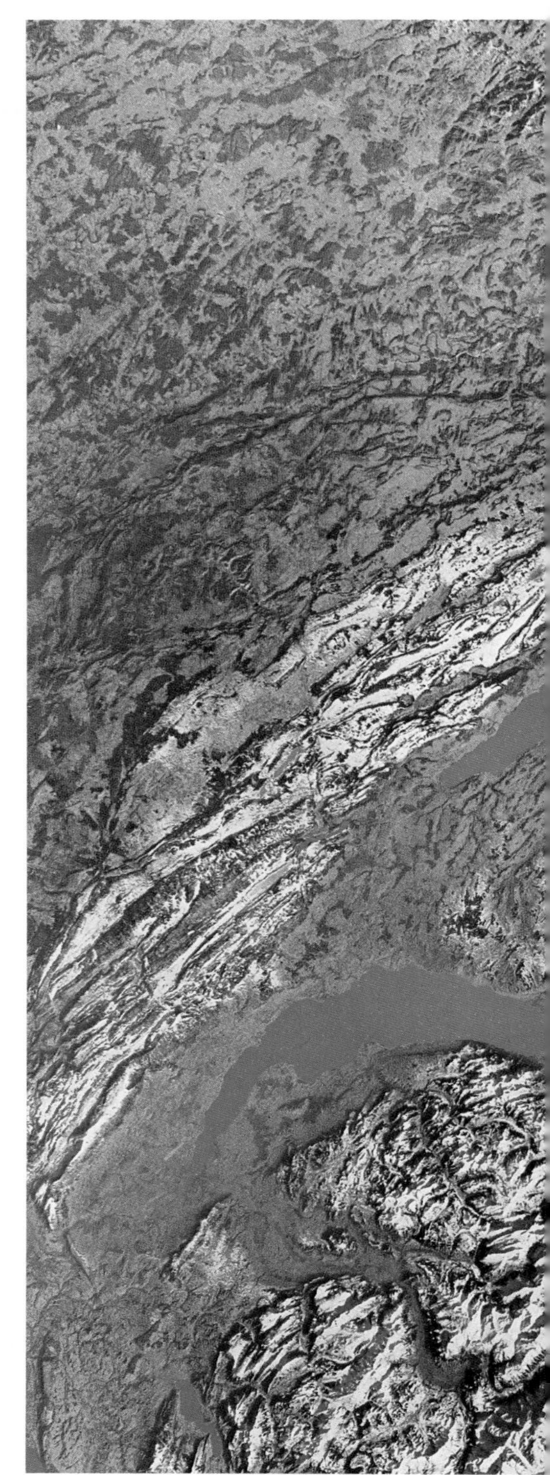

5.31

For example: Grindelwald

The Grindelwald region is a good example of the links between water and energy, and tourism and agriculture, tightly interwoven with a traditional social fabric. The actual climate changes will become the test for the flexibility of the inhabitants.

5.32
Eiger, Mönch and Jungfrau peaks. This famous panorama has become the major trademark for tourism in Switzerland.

5.33 and 5.34
Water and Energy: Springs discharge their minimum during the periods of peak tourism. Highly-developed infrastructures mean that supply nearly matches the demand. Rising temperatures might increase water supply, but perhaps lead to less snowfall and marginal ski conditions. These possibilities force inhabitants to consider artificial snow-making. Water and energy consumption would then rise again.

5.32

5.33

Water Availability and Consumption
Current supply
Supply after global warming
Consumption
October
April
October

5.34

5.35

5.35 and 5.36
Tourism: If the future winter snows become less reliable, especially at Christmas, then ski tourism declines without technical intervention. It is questionable whether an increase in summer tourism could offset the economic loss.

5.37 and 5.38
Agriculture: Traditional methods of cultivation of the parcels in this charming landscape of mixed natural and human imprint can only be maintained thanks to the supplemental income from tourism. If tourism drops, a collapse is in sight without federal subsidies. Is subsidized landscape gardening a goal?

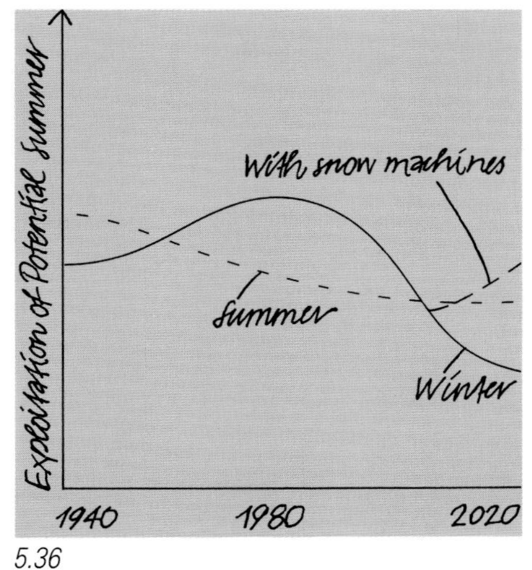

5.36

5.37

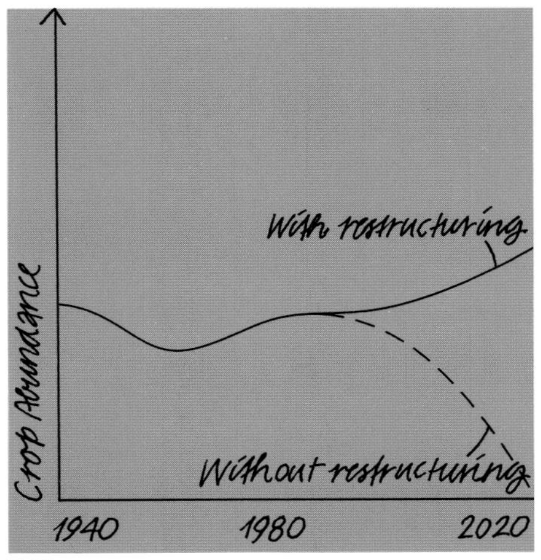

5.38

5 Climate research

The scientific assessment of the IPCC summarizes the state-of-the-art knowledge of climate change under the categories: what is known with certainty, what is calculated with confidence, what is predicted by current models, and what is the best judgement on matters that are more uncertain. It is certain that the Earth's mean surface temperature will increase due to human activities that release additional greenhouse gases to the atmosphere. Furthermore, current model results predict a decadal warming of 0.3 °C, leading to a 1 °C warming above present values by the year 2025 and a warming of 3 °C before the end of the next century.

The timing and magnitude of changes in temperature, precipitation and soil moisture on regional and local scales are highly uncertain. Key areas of scientific uncertainty are the behaviour of clouds and their radiative properties; the exchange of energy between the ocean and the atmosphere and within the ocean; the quantitative uptake and release of greenhouse gases and their chemical reaction in the atmosphere; the mass balance and behaviour of polar ice sheets; and feedbacks between vegetation and climate.

The International Geosphere Biosphere Program, IGBP, and the World Climate Research Program, WRCP, are the two major research programs formulated to provide the data that are needed to reduce the major uncertainties in predicting decadal to century timescale global changes. The WRCP addresses physical aspects of the climate system, whereas the IGBP focuses on biogeochemical interactions in the Earth system leading to global change. Research programs planned under the WCRP/IGBP umbrella include the International Satellite Cloud Climatology Project, the Global Energy and Water Cycle Experiment, the World Ocean Circulation Experiment, the study of Tropical Oceans and Global Atmosphere, the International Global Atmospheric Chemistry Project, and PAGES, a study of past global changes. Both programs are international in character, but causes and effects of climate change touch national boundaries. Therefore, national research programs with similar structures will strengthen the global web of research effort.

Along with improved capacities for observation and computing, progress in these research programs will enable scientists to construct more realistic climate models. At the moment, many interactive processes are modelled separately, and then integrated into larger models as simplified parameterizations, rather than as actual process models. Thus representations of processes like land surface hydrology are plagued by time constant and scaling problems.

The current spatial resolution of large climate models reduces several hundred kilometres into a single data point. The Alps, for instance, become an invisible feature to the model. Regional projections from models should be understood, therefore, more as a set of general possibilities than as accurate predictions. Reducing regional uncertainty is a natural task for national climate programs. ProClim, the Swiss National Climate Program, was inaugurated in 1988, partly in response to WRCP and IGBP research needs. In the ABCs of the ProClim research strategy, A represents modelling efforts to weave current understanding of the alpine chain into the fabric of global climate models; B is a focus on reconstructing rates and impacts of past changes and C – projects include integration of climate and ecosystem parameters at a series of critical threshold sites in the Alps. Can we define criteria for the potential response of ecosystems to specific climate scenarios?

Because of the complex nature of the Earth system, the public must be aware that rapid results cannot be expected from these research programs. However, this is not a license for public complacency; it is a warning that flexible strategies for coping with anticipated global changes must be developed, strategies that take uncertainties, possible surprises and the best current understanding of the Earth system into consideration.

Climate – our future? A vision.

6

"Unpleasant suprises in the greenhouse?" This warning comes from one of the early prophets of climate change, Wallace S. Broecker. From his perspective, we are playing Russian Roulette with climate and no one knows what lies in the active chamber of the gun. Ice and sediment cores teach us how suddenly the Earth system can respond in abrupt, step-like changes. Thus people may be shaken up by suprises that affect their social, economic and political fabric. Early humans survived the Ice Ages by adapting. Are we, with all of our technical progress, to succumb to a self-induced Warm Age? John F. Kennedy inspired a nation with the vision to return a man safely from the Moon. The impossible became reality, and returned to us a new vision of planet Earth. Shall we now embark on a global journey of sustainable development?

6 The Globe gets warmer

The world will become warmer. Few doubt this trend. IPCC reports a likely 2 °C warming within the next generation; a 4–5 °C increase seems realistic for the end of the next century, if we follow a business-as-usual scenario. We have no comparable archive record for such large changes in the past. We use the best models of global circulation available in order to estimate potential global warming. These models combine the current scientific understanding of climate processes with projected values of further increases of greenhouse gases. At the moment, it is more difficult to guess future trends in greenhouse gases than to model the general behavior of the physical climate system. Business-as-usual is our only valid scenario until real changes in energy usage occur. The behavior of the oceans is one of the great scientific unknowns. We only understand rudimentary feedbacks between atmosphere and ocean, or the role of oceanic circulation on global climate. As far as we know, global circulation of water masses is driven by small differences in salinity and temperature. Salts are enriched in the North Atlantic with respect to the North Pacific waters due to greater evaporation. North Atlantic water masses cool, sink and flow like a giant river along the seafloor, around South Africa, across the Indian Ocean and up into the Pacific Ocean. This salty river from the Atlantic to the Pacific is countered by a flow of less-saline, light surface water masses in the opposite direction. This water mass carries heat towards Europe, and the cycling continues. We have difficulties predicting the potential dangers related to present climate, because the system is self-stabilizing. Ice-core studies suggest that the climate fluctuations of the last Ice Age represent a Flip-Flop mechanism between two basic stable states of climatic response. The last 10 000 years have been remarkably uniform. How will the cycles and the ocean-atmosphere system respond to rising greenhouse gas concentrations?

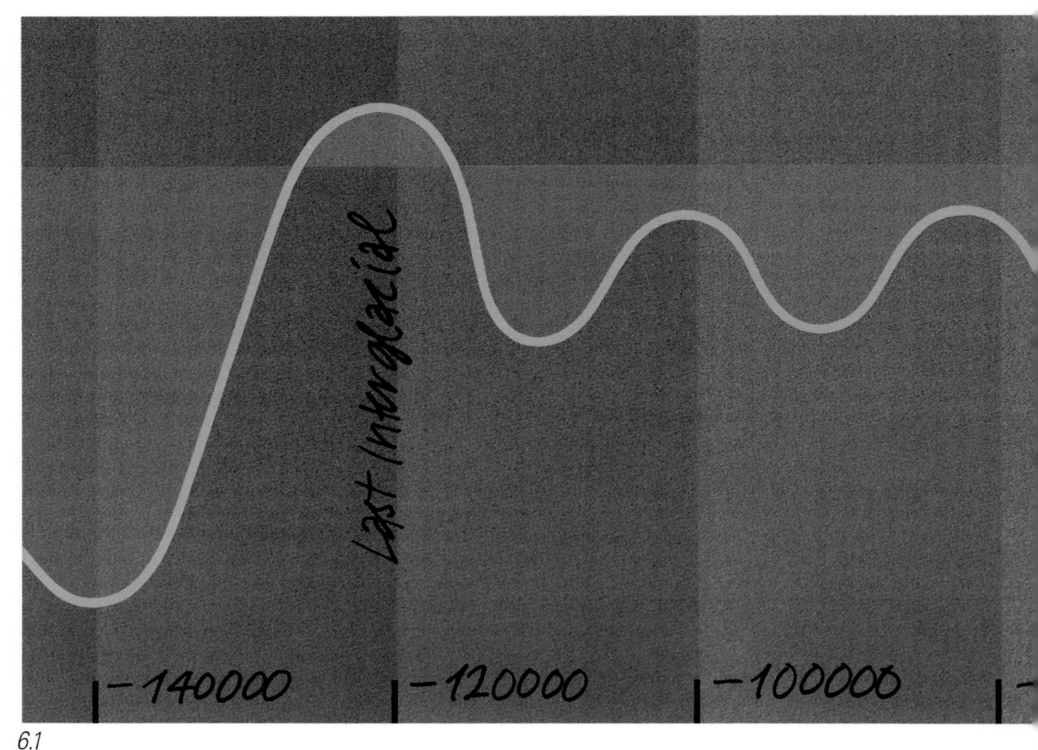

6.1

Estimated curve of temperature changes over the last 140 000 years, and those projected for the self-induced future Warm Age.

6.2

Computer models of the physical climate system confront fundamental barriers. Different parts of the system influence each other in ways that may leave the whole system unpredictable. How might cloud cover change and affect radiation balances? How might this affect the exchange of energy between the oceans and atmosphere? What are the consequences for ocean circulation and weather patterns? Everyday we are confronted with new suprises. We accept these and adapt. These now include unpleasant suprises in the greenhouse.

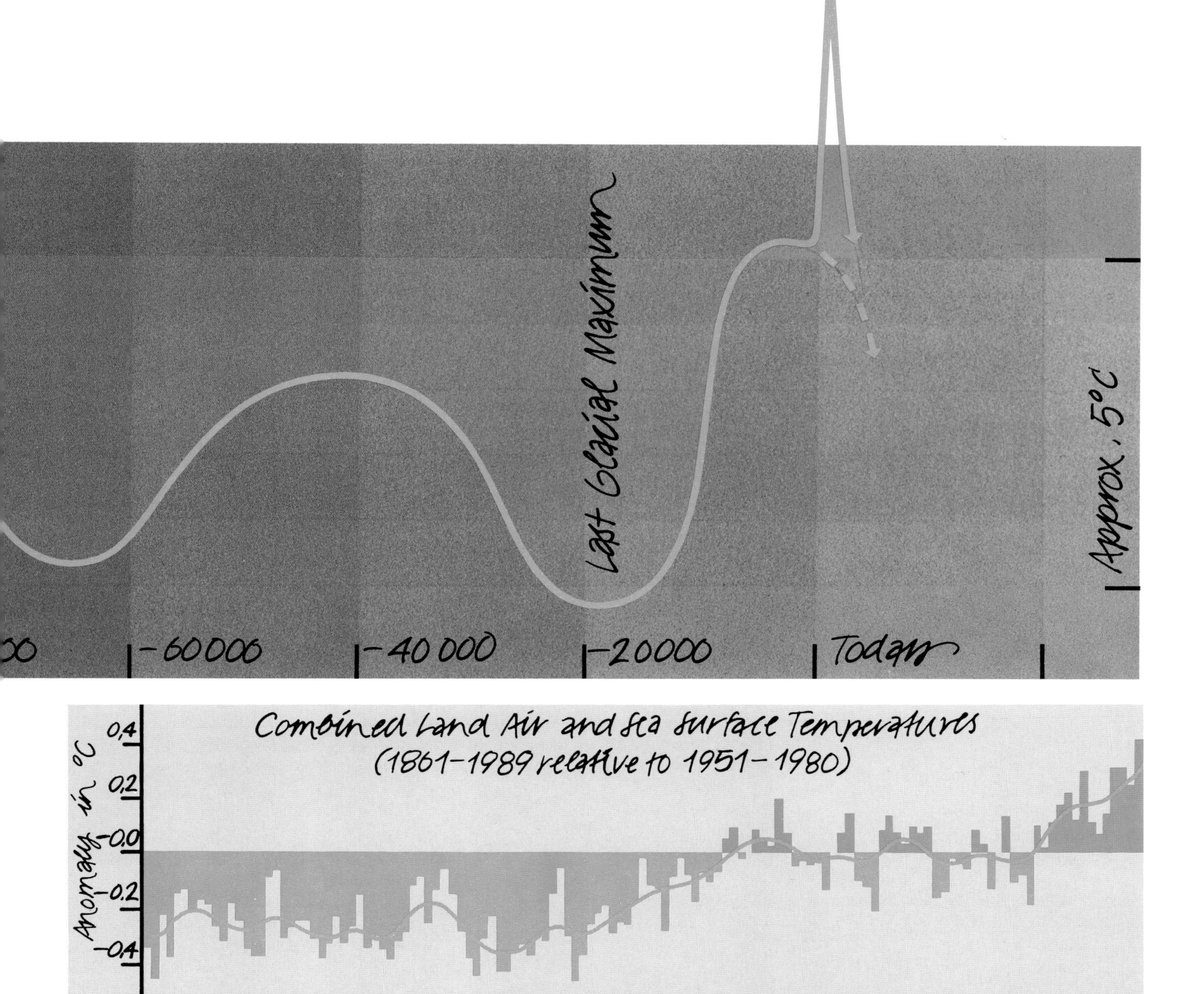

6.3

Deviations of global mean temperatures since 1861, relative to the average of 1951–1980. The ocean delays the direct greenhouse warming. Like a giant pot-belly stove, oceans must first heat up before releasing warmth. They continue to warm the atmosphere long after the fuel burns out.

The seven warmest years of our century are between 1980 and 1990 ranging from 15.25° to 15.45° Celsius for the global average temperature. 1990 is on the top.

6 Sea level rises

The media and, with it, the public are very preoccupied by the scenarios of sea-level rise related to a general warming trend. Subconsciously, the scenarios recall archetypes of biblical flooding. Although measurements are extremely difficult to interpret, records indicate a global rise of about 12±5 cm since 1900. According to the IPCC business-as-usual scenario, simple models predict future sea-level rise of 6 cm per decade. The mean global rise could add up to 65 cm by the end of the next century, but will vary greatly from region to region. The largest component of the rise derives from the thermal expansion of ocean water. Catastrophic flooding due to a sudden disintegration of the West Antarctic Ice Sheet appears unlikely within the next century. More critical dangers stem from currently unpredictable increases in storm intensity. European coastlines tasted the fury of such increases in early 1990.

6.4
Estimated global rise of sea level. The shorter data set from 1930 to 1980 – open circles – covers more stations.

6.5 and 6.6
The technical battle against the sea to protect land. Halligen Island on the German North Sea coast and colassal dykes in The Netherlands.

6.7
Western European industrial society can respond easier than others, whose clocks are at 5 minutes to midnight.

6.5

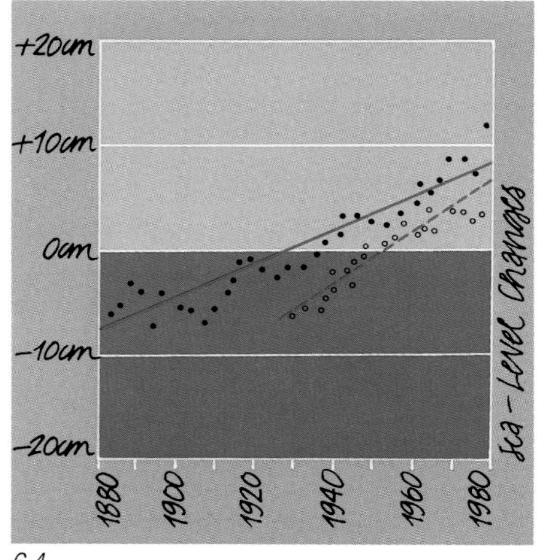

6.4

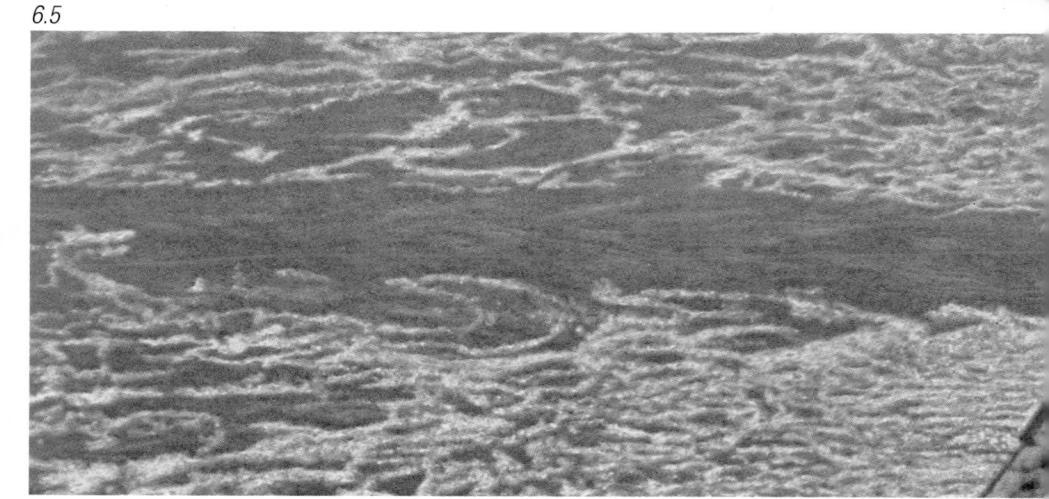

6.6

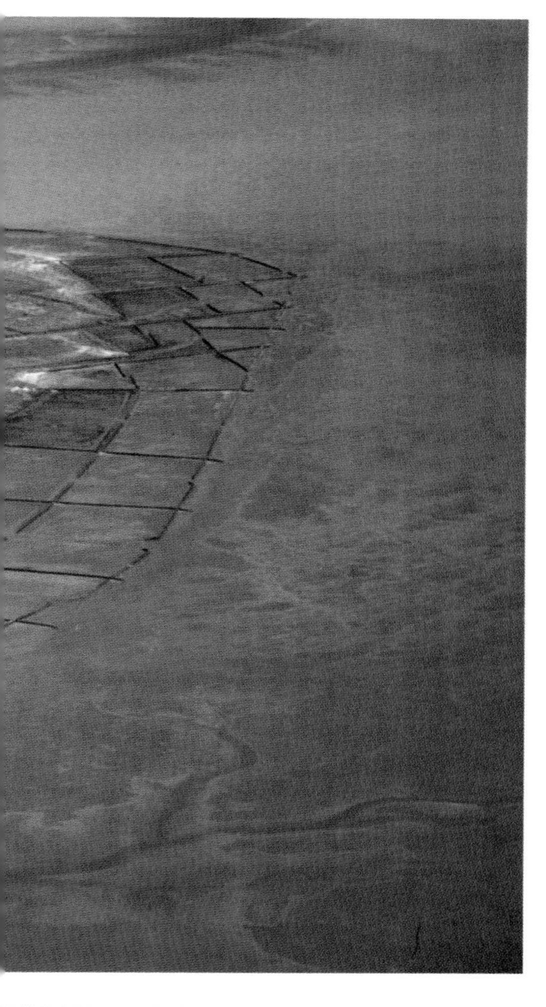

6.7

6 The desert grows?

Shifts in the global hydrological cycle caused by global warming are even less predictible than rises in sea level. There will be winners and losers. Precipitation patterns will not consider the population distribution or technical possibilities. For example, we expect the tropics to show an increase in convective rainfall. This plague of bounty might increase erosion. If climate belts shift northward, the problems of the Sahel Zone will defuse, but ignite somewhere else. The dangers of desert encroachment are present in many regions of the world, in combination with overusage. Mitigating measures are difficult, because forecasts from computer models of global climate change are inadequate to identify which local areas will need help first. The race against time may be lost without a concerted focus on more research. Scientists in developing countries must be encouraged in their efforts to study the sensitivity of local environments. New research partnerships will strengthen the global web.

6.8
Will the deserts continue to grow?

6.9 and 6.10
Wheat fields in the Midwest of North America . . . and visions of their future landscape if the predicted warming arrives.

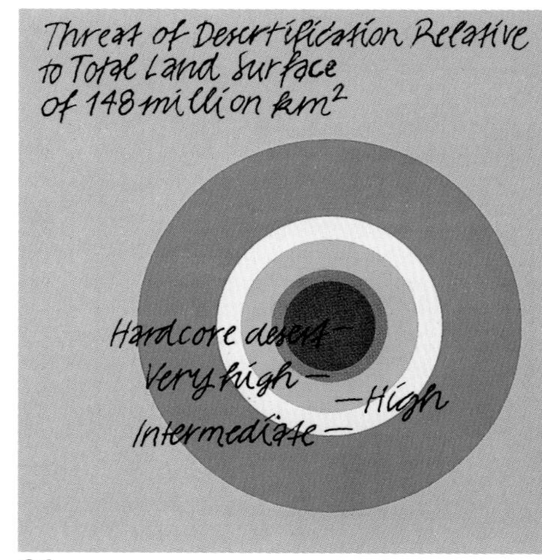

6.8

6.9

6.10

6 Migration increases?

Today, there are some 10 to 15 million refugees. Half are children. Most live in developing countries. According to the Geneva Convention of 1967, a refugee is a human who cannot return to his/her country because there is a *reasonable fear that his association with a particular race, religion, nationality, political movement or social group will lead to persecution.* Most people still flee from political unrest, but global change may drive future waves of migration. Shifts of climatic belts would have the greatest affect on those countries which already fight a battle of survival: the countries pressured by population growth. If their livelihood is threatened, people move. According to the UN Charter, every person has a right to food, shelter, work, and education. Who has a right to prevent migration?

6.11

Most refugees flee from one poor country to another…

6.12 and 6.13

He can't stay here; there, he feels like an outsider. What is the solution to his future?

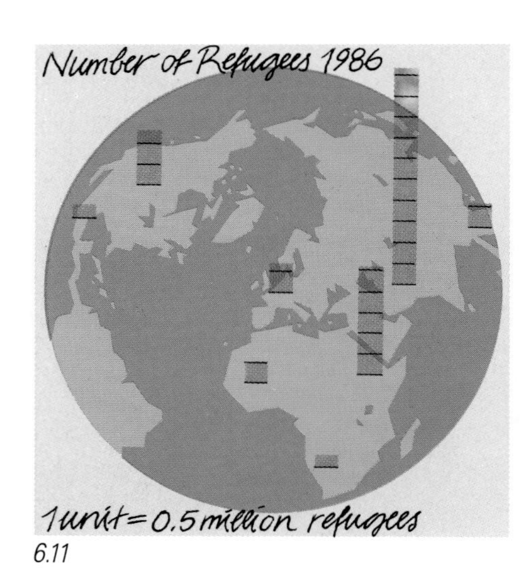

6.11

6.12

6.13

6

6.14

6.15

6.17

Everything flows...

We live within the river of time and seek to grasp the moment. Tradition provides the solid yardstick of relative changes in our lives. We bend to tradition, as we have bent to climate in the past. During a lifetime, we feel tradition as something constant. However, tradition also bends, albeit in slowmotion.

Will changes accelerate with climate dynamics?

If only a portion of the climate predictions described in this book come to pass within the next one or two generations, they will fundamentally change our natural environment. The magnitude of changes will not fail to alter custom and culture, even if we try to preserve them by artificial means. White Xmas in Australia?

6.14
Tradition on plastic chairs in Valais, Switzerland. Signs of change.

6.15
Swiss farmhouse. Will traditional garlands of geraniums be supplanted by oleander and cypress?

6.16
Japanese gardens as a tradition. Will landscaping help us retain familiar environments?

6.17 to 6.19
The foreseeable future hardly requires camels at the Swiss Federal Parliament. The backdrop of Bernese hills only erodes at geologic rates. Still, Nature will print with a new stamp.

6.16

6.18

6.19

6

The Farmer as a mountain gardener...

How might climate changes affect the Alps? Large differences in altitude demand a large palette of scenarios. In all honesty, the research effort to provide data and criteria for evaluation is just at the doorstep. The formulation of long-term strategies requires a clear political and financial concensus. Scientists cannot yet provide the proof requested by today's policy makers. Are worst-case-scenarios the only way to start?

In many parts of the Alps, tourism and agricultural economies are potentially most threatened by change. A region not only lives from- but also through- the traditional agricultural customs, which have functionally sculpted the landscape and provided charm and attraction for tourism. Traditional customs could become frozen into mere folklore rather than following a natural function as pivots in a rapidly changing environment. Realities may pile-up against folklore, eventually leading to total breaks with the past. Can new identities be found in an artificially modern realm?

6.20 to 6.22

Harvesting hay in the Alps: 1960 and 1980... and perhaps in the year 2000?

6.20

6.21

6.22

6.23

...palm or Cypress trees in the Alps?

This is an unlikely scenario. Still, if winters in central Europe follow the trends of recent years, tourism is in for major structural adaptations. If the region experiences 3 °C warming, most glaciers will disappear in the 21st century. Future tourists will be awed by a new face for the Alps. One thing is certain: change will come. Thus we must take precautions to avoid drowning in a flood of complacency.

6.23
The Schreckhorn in the Bernese Alps, behind a retouched landscape.

6

What can We do?

The Second World Climate Conference did not produce recommendations for a global plan of action. They did agree of the science. Planning for action was shifted to UNCED 1992, the United Nations Conference on Environment and Development, following the motto "wait and see". A climate convention certainly cannot provide patent answers for everyone, everywhere. Stabilizing the greenhouse gases at present atmospheric concentrations requires a more than 60% *immediate* reduction in CO_2, CFC's and nitrous oxide emissions and a 20% reduction in methane. Even if immediate seems unrealistic, the goal of stabilization by the middle of the next century can only be achieved if we start now. Innumerable possibilities exist for producing and applying energy more efficiently. Energy conservation is not a priority when energy is too cheap. Renewable and CO_2 free energy production technologies are wide spread available, but their use lacks political and financial support. Development of new alternative energy resources must be fed with the real fuel-of-motion, money. A continuing obstacle to the development of alternative energies is that assessments of economic viability fail to include environmental impacts among the costs of burning fossil fuels. Some of our most precious resources are free, for the moment. If we do choose to conserve our resources, price tags must be set at levels of pain which stimulate market processes. Success depends on a balance among society, environment and technology.

6.24

Humans produced 24 billion tons of CO_2 in 1990. This represents over 4 tons per capita per year. The goal for 2050 is one ton per capita per year, for our own good...

6.25

...Who can master the crisis: reason, Nature, or both?

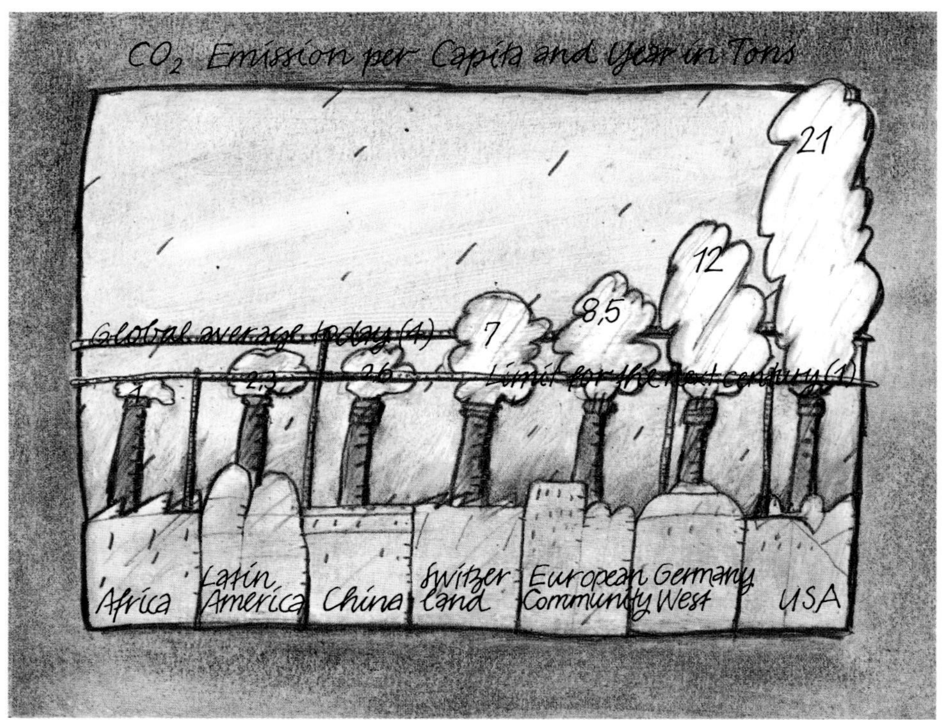

6.24

6.25

If society takes scientific warnings on further greenhouse gas increases seriously, active steps must be taken. The 1988 Toronto Conference on the Changing Atmosphere made a series of moderate recommendations: a reduction of 1.5 percent per year in CO_2 emissions should lead to a 20 percent reduction in emissions by 2005 and then to 50 percent reduction by the middle of the 21st century. This burden must mainly be carried by the rich, which is logical, considering that about 20 percent of the people now consume more than two thirds of the world's primary energy. One year later, the 1989 World Energy Conference in Montreal delivered the sobering reality. Even very modest scenarios of energy consumption showed a distinct upward trend. In 1990, we count the number of countries who have volunteered to reduce CO_2 emissions on one hand. The world's largest consumers are not among these fingers.

6.26
Future trends of CO_2 emissions between desires and reality. If more hands are not offered to solve the problem, by 2020 the level of desired reduction will effectively be doubled in a business-as-usual world.

6.27
Energy is a precious resource. Our future course will be set by how we handle it.

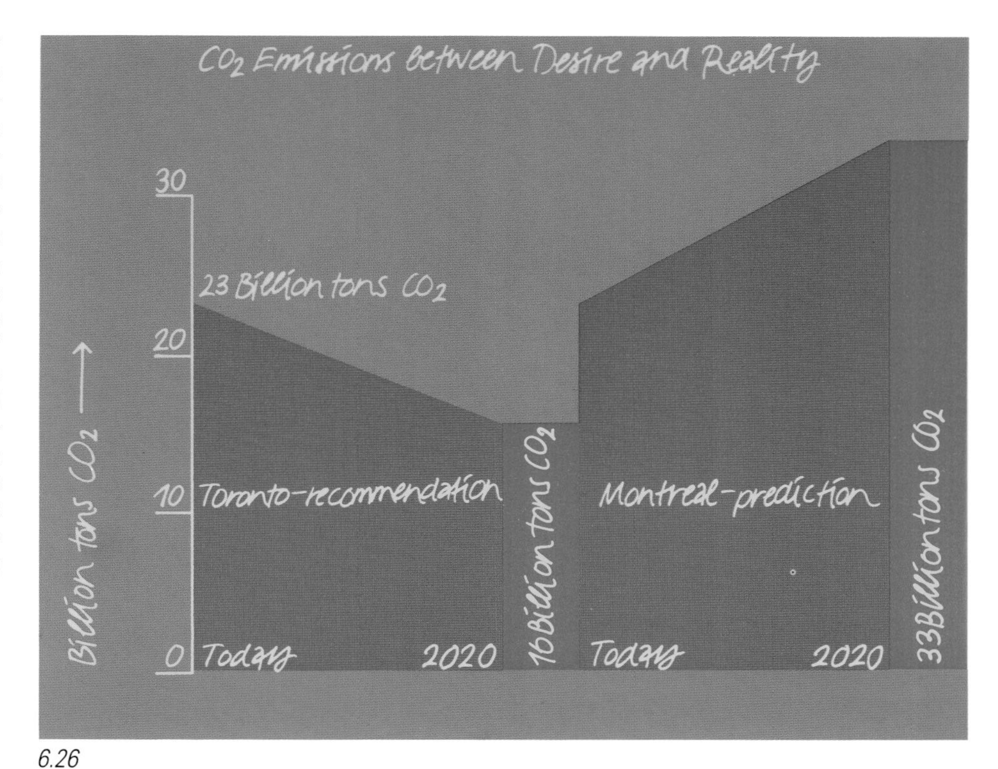

6.26

6.27

6 Children, your future

The world continues to change at an ever-increasing tempo. That which exists is gone tomorrow. Fashions fade. Yesterday's technical fantasies become boring realities. Instant communication leads to a video-clip awareness of a seemingly exponential flood of events. Climate adds another dimension. Climate can change our environment at the same pace that we change climate. In order to rise above a doomsday mentality and encourage positive modes of behavior and thought, we need more pensive reflection. Our future is in the hands of our childern. Their lives will be moulded by change and adaptation. Reaching outward to understand the world provides a firm handle for inner harmony. Real knowledge brings strength. Ignorance, fear and apathy in an endangered world lead down deadend streets. A belief in the vision of a sustainable world should accompany our collective steps into the next century.

6.28 to 6.30

School kids from the International School of Berne, Switzerland: What's in store for them?

6.28

6.29

6

CLIMATE OUR FUTURE
it's hot, hot! help! HELP!!
KLIMA UNSERE ZUKUNFT?
CLIMATE OUR FUTURE
HET KLIMAAT ONZE TOEKOMST
CAFE CENTRAL
The force shield protects the city

6.30

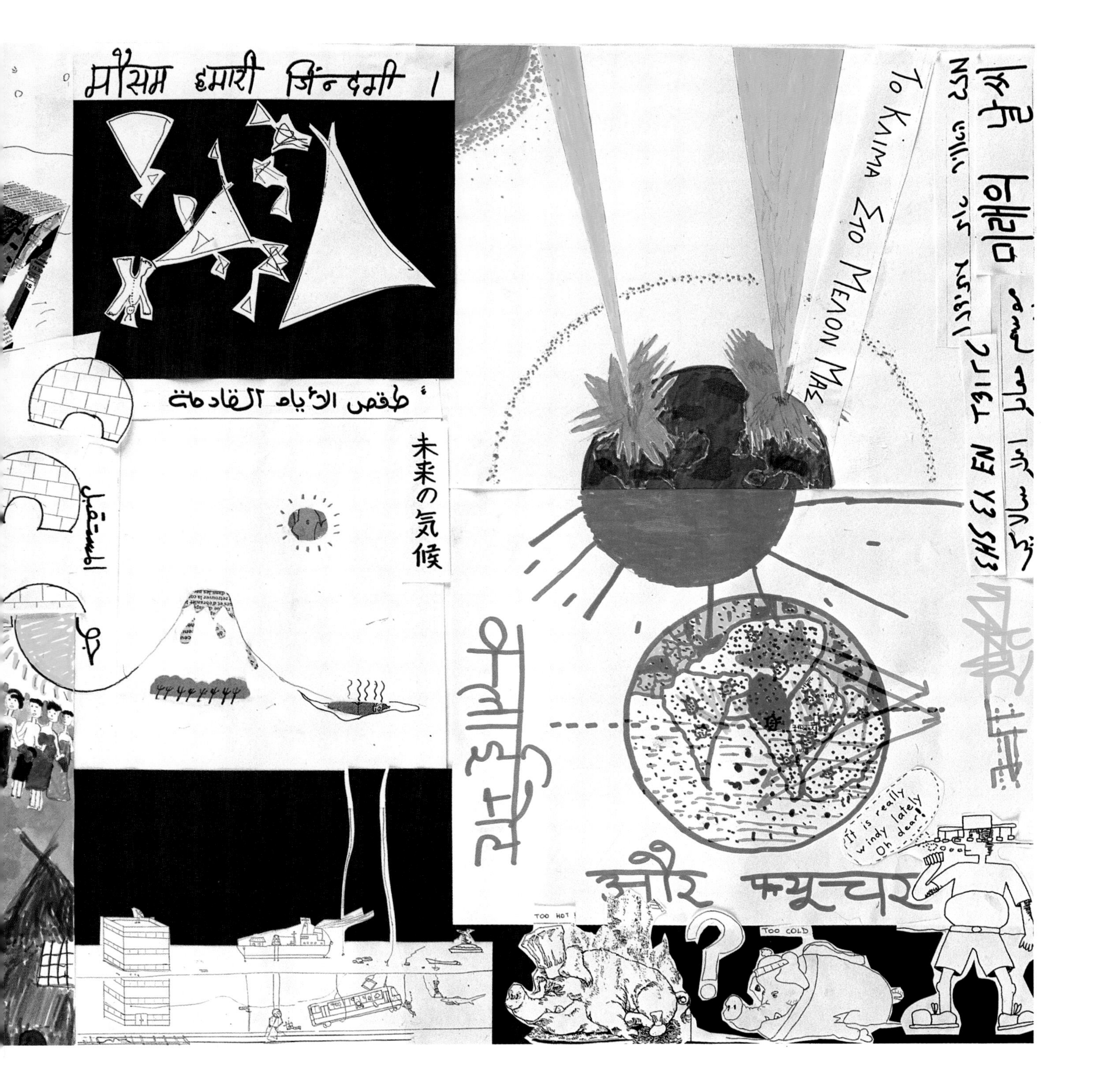

मौसम हमारी जिंन्दगी ।
طقس الأيام القادمة
未来の気候
It is really windy lately Oh dear!
और फ्यूचर
TOO HOT!
TOO COLD

6 Climate – our future? A vision

We are all sitting in the same boat. That boat is our planet Earth. The next decades will show whether we can master the global challenges of environmental change and population growth together.

Human causes of change are closely linked to world population and energy use. The rising atmospheric concentration of greenhouse gases is a prime example. To stabilize greenhouse gas concentrations at today's levels would require immediate 60–80% reductions in emissions of the long-lived gases CO_2, N_2O and CFC's. Even moderate measures, such as reducing 1988 CO_2 emissions by 20% by the year 2005, seem unattainable in light of industry and policy obstacles that block the path to reductions. And all scenarios, regardless of the severity of emissions reductions, nevertheless show increasing greenhouse gas trends in the coming decades. Will the man in the street take warnings from the scientific community seriously, or will he merely wait and see?

Taking steps to allieviate the enhanced greenhouse effect is the responsibility of all nations – both industrialized and developing. Any national technological or policy decision will affect the timing and severity of global warming. However, since less than one third of the world population today consumes more than three quarters of the energy produced, it is clear that the industrialized countries must take the lead.

Technology transfer and financial support to developing countries will strengthen the measures of these countries. Technologies related to energy production and consumption that can reduce human impact exist. Increased energy efficiency and conservation must be paralleled by a transition from the use of carbon-intensive to carbon-free fuels. Options range from advanced nuclear technologies to biomass recycling and solar hydrogen fuel.

An end to the use of CFC's is on the way. But, it took nearly 20 years from the first scientific warnings that CFC's might be harmful for policy makers to agree on a concrete plan of action. We cannot wait that long to develop policies for reducing emissions of other greenhouse gases. An immediate reduction in CO_2 emissions is needed to curb the enhanced greenhouse warming of our planet. Steps to halt further deforestation, to begin reforestation projects, to develop non-carbon fuel sources, and to impose necessary but politically unpopular policies such as substantially higher taxes on fossil fuels must be taken now. Because the rapid increase in world population will be a major factor in determining the future course of greenhouse gas emissions, it is essential that global climate change strategies must also include measures to reduce rates of population growth.

The data are in. National political leadership must now face the challenge of developing policies to respond to the regional and local impacts of predicted global changes, in the light of inherent uncertainties. Will we embark on a global journey of sustainable development?

Three things
influence human thought:
climate, politics and religion.
Voltaire 1756

6 Credits

1. Weather and climate

Title page: Andreas Stettler
Photographs:
Dipartimento dell'Ambiente, Bellinzona (1.28)
ESA, Darmstadt (1.3, 1.35)
Karl-Heinz Hack (1.5)
Anna Holström (1.49)
Alain Jeanneret (1.21)
Giovanni Kappenberger (1.18, 1.19, 1.26)
Beat Käslin (1.9, 1.13, 1.15)
Heinrich Rufli (1.38, 1.51)
Erich Schneiter (1.2, 1.7)
Ulrich Schotterer (1.10, 1,39, 1.52, 1.53, 1.56, 1.59)
University of Berne, Institute of Geography (1.31, 1.33, 1.36, 1.37)
Klaus Wernicke (1.30)
Matthias Winiger (1.44, 1.45)
Anne Zwahlen (1.47)
Graphics and illustrations:
Verena Baumann (1.25, 1.27, 1.46, 1.55)
Elsi Brönnimann (1.29, 1.42, 1.48)
Silvia Brühlhard (1.1, 1.4, 1.8, 1.11, 1.14, 1.17, 1.20, 1.23, 1.40, 1.41, 1.43, 1.57, 1.58)
Walter Buri (1.6, 1.12, 1.16)
Lukas Machata (1.43)
Roberto Renfer (1.24, 1.50)
Andreas Stettler (1.32, 1.34)
Agnes Weber (1.22, 1.54)

2. What we know about climate

Title page: Silvia Brühlhardt
Photographs:
Brigitta Ammann (2.46)
Central library, Zürich (2.50)
Ueli Eicher (2.57)
Niklaus Flüeler (2.67)
Burkhard Frenzel (2.52)
Claus Fröhlich (2.6, 2.9)
Gerhard Furrer (2.32)
Oswald Heer, Urwelt der Schweiz Zurich, published 1865 (2.53, 2.54)
Hanspeter Holzhauser (2.1, 2.42, 2.50)
Kerry Kelts (2.36)
Ernst Kopp (2.17)
Viktor Maurin (2.22)
NASA Washington (2.10, 2.30)
Fritz Röthlisberger (2.40)
Heinrich Rufli (2.3, 2.56)
Ulrich Schotterer (2.2, 2.18, 2.38, 2.39, 2.41, 2.43, 2.71)
Emil Schulthess (2.15)
Jakob Schwander (2.37)
Fritz Schweingruber (2.4, 2.5, 2.20, 2.34, 2.45, 2.47)
Universiy of Berne, Institute of Physics, (2.8)
Carl Zeiss AG, Switzerland (2.16)
Heinz Zumbühl (2.61, 2.62, 2.63, 2.64, 2.65, 2.66)
Graphics and illustrations:
Verena Baumann (2.19, 2.31, 2.33)
Walter Buri (2.59, 2.60)
Elsi Brönnimann (2.26, 2.27, 2.28, 2.29, 2.58, 2.66, 2.68, 2.70)
Silvia Brühlhardt (2.49, 2.55)
Luke Machata (2.11, 2.12, 2.13, 2.14, 2.23, 2.35, 2.44, 2.51)
Roberto Renfer (2.7, 2.9, 2.19, 2.21, 2.25)
Andreas Stettler (2.24, 2.69)

3. Climate, humans and landscape

Title page: Elsi Brönnimann
Photographs:
Alpine Museum, Berne (3.43, 3.45)
Georges Grosjean (3.40)
Hanspeter Holzhauser (3.31)
Landesdenkmalamt Baden-Württemberg, Karlsruhe (3.42)
Bruno Messerli, Universität Berne (3.14, 3.15, 3.29)
Museum of Ethnography, Basel (3.6)
Museum of Fine Arts, Basel (3.41)
Museum of Fine Arts, Berne (3.48)
Museum of History, Berne (3.34)
National Museum, Kopenhagen (3.6)
Nebelspalter, Rorschach 1889 (3.46)
Martin Obrist (3.28)
Ulrich Schotterer (3.1, 3.2, 3.12, 3.17, 3.18, 3.20, 3.37)
Schuler Verlag, Stuttgart (3.39)
United Nations 164645, John Isaac (3.27)
Walservereinigung, Chur (3.35)
Heinz Zumbühl (3.22, 3.23, 3.36, 3.38)
Graphics and illustrations:
Verena Baumann (3.11, 3.13, 3.16, 3.19, 3.21, 3.24, 3.49)
Elsi Brönnimann (3.4)
Silvia Brühlhardt (3.30, 3.32, 3.46, 3.47)
Walter Buri (3.26)
Edith Helfer (3.25)
Luke Machata (3.3, 3.33, 3.44, 3.50, 3.51)
Andreas Stettler (3.49)
Agnes Weber (3.7, 3.8, 3.9, 3.10)

4. People – climate

Title page: Luke Machata
Photographs:
Werner Berner (4.14)
CIRIC, Lausanne (4.3)
Jacques E. Cuche (4.8)
Catherine Graf (4.3)
Toni Linder (4.16)
Luc Meylan (4.7)
Heinrich Rufli (4.12)
Ulrich Schotterer (4.9, 4.22, 4.29)
Fritz Schweingruber (4.16)
Hans Turner (4.25)
Matthias Winiger (4.4, 4.5, 4.6, 4.27)
Graphics and illustrations:
Eva Baumann (4.17)
Verena Baumann (4.15, 4.26, 4.28)
Elsi Brönnimann (4.11, 4.13, 4.30, 4.32)
Walter Buri (4.10, 4.18, 4.19, 4.20, 4.25)
Catherine Eigenmann (4.14)
Luke Machata (4.1, 4.2, 4.15, 4.21, 4.23)
Andreas Stettler (4.31)

5. Climate research

Title page: Walter Buri
Photographs:
Alpine Museum, Berne (5.5)
Dee Breger (5.14)
ETH Zürich Institute of Physics (5.13)
Beat Käslin, (5.34)
Toni Linder (5.24)
Philippe Plailly (5.8)
Ulrich Schotterer (5.9, 5.10, 5.11, 5.12, 5.22, 5.26, 5.28)
Klaus Seidel (5.31)
SMA Zürich (5.3, 5.4, 5.20)
University of Berne Institute of Geography (5.7, 5.18)
University library Neuchatel (5.1)
Urs Wiesmann (5.33, 5.35, 5.37)
Graphics and illustrations:
Eva Baumann (5.34, 5.36, 5.38)
Walter Buri (5.27, 5.28, 5.32)
Catherine Eigenmann (5.2, 5.23)
Katja Leudolph (5.19, 5.21)

Luke Machata (5.25, 5.30)
Roberto Renfer (5.7, 5.15, 5.16, 5.17)
Andreas Stettler (5.9, 5.29)
Karin Widmer (5.6)

6. Climate – our future? A vision

Title page: Roberto Renfer
Photographs:
Georg Budmiger (6.20, 6.21)
Embassy of The Netherlands, Berne (6.6)
Ulrich Schotterer (6.14, 6.15)
Emil Schulthess (6.16)
United Nations, 129998, Jerry Frank (6.12)
Klaus Wernicke (6.5)
Matthias Winiger (6.9)
Graphics and illustrations:
Verena Baumann (6.10)
Elsi Brönnimann (6.29, 6.32)
Silvia Brühlhardt (6.17, 6.18, 6.19, 6.27)
Walter Buri (6.2, 6.3, 6.22, 6.26,)
International School of Berne (6.28, 6.29, 6.30)
Luke Machata (6.11, 6.13, 6.24, 6.25)
Sibylle von May (6.4, 6.8)
Roberto Renfer (6.1, 6.23)
Andreas Stettler (6.7)

Recommended for further reading

John Imbrie and Katherine P. Imbrie
Ice Ages–Solving the Mystery. Enslow Publishers, New Jersey, 1979.
Wallace S. Broecker
How to Build a Habitable Planet. Eldigio Press, Palisades, New York, 1985.
Bert Bolin, Bo R. Döös, Jill Jäger and Richard A. Warrick
SCOPE 29: The Greenhouse Effect: Climatic Change and Ecosystems. John Wiley & Sons, New York, 1986.
Stephen H. Schneider
Global Warming: Are We Entering the Greenhouse Century? Sierra Club, San Francisco, 1989.
Dean Edwin Abrahamso, editor
The Challenge of Global Warming. Natural Resources Defense Council, Island Press, Washington, DC, 1989.
John F. Mitchell
The Greenhouse Effect and Climate Change. Reviews of Geophysics, 1989.
Johnathan Weiner
The Next One Hundred Years, Shaping the Fate of Our Living Earth. Bantam Books, New York, 1990.
Enquete Kommission of the German Bundestag, editor
Protecting the Earth's Atmosphere: An international Challenge. D–5300, Bonn 1, Bundeshaus, 1990.
World Resources 1990–1991
A Guide to the Global Environment. Oxford University Press, New York. 1990.
IGBP: A Study of Global Change
Global Change Report No. 12 IGBP Secretariat, Royal Swedish Academy of Sciences, Stockholm, 1990.
Intergovernmental Panel on Climate Change, The IPCC Scientific Assessement. Edited by H. T. Houghton, G. J. Jenkins and J. J. Ephraums. Cambridge University Press. Cambridge, 1990.
Climate Change: Science, Impact and Policy
Proceedings of the Second World Climate Conference. Ed. by J. Jäger and H. L. Ferguson. Cambridge University Press. Cambridge 1991.

Book credits

Cover and chapter summary illustrations by Peter Andermatt; handlettering by Andreas Stettler